Effective Teamwork

Ten Steps for Technical Professions

DAVID L. GOETSCH
CEO, Institute for Continual Improvement

NetEffect Series

Upper Saddle River, New Jersey
Columbus, Ohio

Library of Congress Cataloging-in-Publication Data

Goetsch, David L.
Effective teamwork: ten steps for technical professions/David L. Goetsch
p. cm. - (NetEffect series)
Includes bibliographical references and index.
ISBN 0-13-048527-6
1. Teams in the workplace. 2. High technology industries. 3. Project management. I. Title. II. Series

HD66.G63 2004
658.4'02-dc21

2003013516

Editor in Chief: Stephen Helba
Executive Editor: Debbie Yarnell
Editorial Assistant: Jonathan Tenthoff
Production Editor: Louise N. Sette
Production Supervision: Gay Pauley, Holcomb Hathaway
Design Coordinator: Diane Ernsberger
Cover Designer: Ali Mohrman
Production Manager: Brian Fox
Marketing Manager: Jimmy Stephens

This book was set in Goudy by Carlisle Communications, Ltd. It was printed and bound by R.R. Donnelley & Sons Company. The cover was printed by Coral Graphic Services, Inc.

Pearson Education Ltd.
Pearson Education Singapore Pte. Ltd.
Pearson Education Canada, Ltd.
Pearson Education—Japan
Pearson Education Australia Pty. Limited
Pearson Education North Asia Ltd.
Pearson Educación de Mexico, S.A. de C.V.
Pearson Education Malaysia Pte. Ltd.

10 9 8 7 6 5 4 3 2 1
ISBN 0-13-048527-6

Contents

About the Author

David L. Goetsch is provost of the joint campus of the University of West Florida and Okaloosa-Walton Community College and professor of management, quality, and safety. Dr. Goetsch is also president and CEO of the Institute for Continual Improvement, a private consulting firm dedicated to the continual improvement of employees, organizations, and communities. Dr. Goetsch welcomes feedback from his readers and may be reached at the following email address: ddsg2001@cox.net.

ACKNOWLEDGMENTS

The author thanks Savannah Goetsch and Caleb Sutton for their assistance with editing and proofreading the original manuscript.

Introduction

Effective Teamwork— A Ten-Step Model

No member of a crew is praised for the rugged individuality of his rowing.

—Ralph Waldo Emerson

TEAMWORK: THE RATIONALE

Effective teamwork is essential to the success of technology companies in today's competitive global marketplace. There are two reasons for this. First, the work of such companies is by its very nature team oriented. Projects are undertaken by teams consisting of engineers, technicians, and other technical personnel.

Second, a team whose members work well together can outperform even the most talented collection of individuals whose efforts are not mutually supportive. This is because a well-trained team is more than just the collective talents of its individual members: The whole is more than just the sum of the parts. Effective teamwork tends to multiply the performance of its members by enhancing their strengths and compensating for their weaknesses. One sees this phenomenon over and over in the world of sports. Two tennis players, neither of whom could win a singles championship, are unbeatable as a doubles team. Five basketball players, not one of whom is an MVP-caliber player, together make a championship team.

Coaches often speak of the "intangibles" that lead teams to perform at higher levels than would be expected of their individual members. In reality, these so-called intangibles are anything but intangible. Rather, they are predictable factors that can be developed and nurtured. They include the following:

1. Effective teams provide a vehicle for taking advantage of the complementary strengths of individual members (the strengths of one member complement and magnify the strengths of other members).
2. Effective teams provide a vehicle for taking advantage of the compensatory strengths of individual members (the specific strengths of one member compensate for the specific weaknesses of another member).
3. Effective teams satisfy the basic human need for association, the need to belong to a group of which the individual can be an important member.
4. Effective teams develop an *espirit de corps* and a level of trust that make the members want to perform well for each other.
5. Effective teams satisfy the basic human need of individuals to gain the respect and esteem of a peer group.
6. Effective teams provide a vehicle for the discovery, development, and application of leadership skills among members.

WHAT IS A TEAM?

A team is a group of people who work together toward a common goal. The common-goal aspect of teams is critical, a point that is evident in the performance of athletic teams. A basketball team in which one player hogs the ball, plays the role of prima donna, and pursues her own personal goals (a personal high-point total, MVP status, publicity, and so on) rarely wins against a team in which all players pull together toward the common goal of winning the game.

An example of teamwork succeeding over individualism is the defensive strategy employed by the Miami Dolphins of the National Football League (NFL) during the early years of the franchise. In a sport that encourages and promotes media stars, no single member of the Dolphins' defense stood out above the others. In fact, although it was arguably the best defense in the NFL at that time, individual members of the team were not well known—hence the nickname "no name defense." Taken separately, the individual team members were less impressive than many of their big-name counterparts on other teams. Bigger defensive ends, stronger tackles, quicker linebackers, and swifter defensive backs played on competing teams. But the Dolphins' defense excelled by working together as a team.

TYPES OF TEAMS

Three types of teams are commonly used in technology companies: work teams, improvement teams, and standing committees. The similarities of and differences among them are explained in the following paragraphs.

Work Teams

Work teams are teams that do the normal, everyday work of technology companies. In an engineering firm, for example, a project might be assigned to a work team consisting of a project manager and several engineers, CAD technicians, and clerical support personnel. The individuals in each of these categories work together as a team to complete the engineering project. This engineering team might pass its completed design along to a manufacturing team, so that a product can be produced, or to a construction team, so that a structure can be built.

Improvement Teams

Improvement teams are typically cross-functional teams assigned the task of improving a given process or some specific function of the organization. The term "cross-functional" means that the team is composed of representatives from a variety of functional areas or departments within the organization. Improvement teams can be temporary or permanent. Temporary teams are assigned one specific improvement project and disbanded once their assignment is completed. An engineering firm might want to improve the process it uses for developing prototypes. To achieve this goal, it might form a temporary or ad hoc team with representatives from all functional units having anything to do with producing prototypes. Once the temporary improvement team completes its assignment, it would be disbanded.

A permanent improvement team also consists of representatives from various functional units, but its mandate is continual improvement of processes and functions. Such teams typically decide what processes and functions should be improved and charter ad hoc teams to carry out specific improvement projects. Permanent improvement teams typically are composed of higher-level personnel than are ad hoc improvement teams. The quality council is an example of a permanent improvement team.

Standing Committees

Standing committees are teams responsible for carrying out ongoing assignments related to specific functions or disciplines. They typically advise higher management about decisions relating to the functional area in which they specialize, and they help keep management up-to-date on the latest information relating to this functional area. For example, many engineering, manufacturing, and construction companies have a safety committee to help keep management informed about the latest laws, regulations, and practices as well as to make recommendations about important decisions concerning safety.

WORKING TOGETHER AS A TEAM

People do not automatically work together well in teams. In fact, a variety of human factors work against effective teamwork. According to Peter Scholtes, some of the critical factors are as follows:[1]

- *Personal identity of team members.* It is natural for people to wonder where they fit into an organization. This tendency applies whether the organization is a company or a team within a company. People worry about being outsiders, getting along with other team members, having a voice, and developing good, trusting relationships with teammates. The work of the team will not proceed effectively until its members feel they fit in.
- *Relationships among team members.* Before people in a group can work together, they have to get to know each other and form relationships. When people know and care about each other, they go to great lengths to be mutually supportive. Time spent helping team members get acquainted and establish common ground is time wisely invested. This is especially important now that the workforce is so diverse. Common ground among team members can no longer be assumed.
- *Identity within the organization.* This factor has two aspects. The first has to do with how the team fits into the organization. Is its mission a high priority for the company? Does the team have the support of higher management? The second aspect of this factor relates to how membership on a team affects relationships with non–team members. This concern is especially important in the case of ad hoc teams, because team members usually want to maintain their existing relationships with fellow employees who are not on the team. They might be concerned that membership on the ad hoc team will be detrimental to their relationships with these other colleagues.

CHARACTER TRAITS THAT PROMOTE EFFECTIVE TEAMWORK

Certain behaviors promote effective teamwork. These positive behaviors arise out of certain character traits that must be developed and nurtured if team members do not adequately possess them. Character traits that lead to positive team behaviors are as follows:[2]

- *Honesty.* To build the level of trust that is foundational to effective teamwork, team members must be honest with each other. Honesty is the cornerstone of trust, and trust is the cornerstone of effective teamwork.
- *Selflessness.* Selfless people are willing to put the team's interests above their own. Team members who use the team to advance their own agendas undermine effective teamwork.

- *Dependability.* Team members are mutually dependent. The performance of each individual depends to some extent on the work of the other team members. Consequently, it is important that team members know they can count on each other.
- *Enthusiasm.* The concept of team spirit is real. People who are enthusiastic about their work typically do it better. The good news is that enthusiasm is contagious. Enthusiasm helps team members persevere when the going gets tough.
- *Responsibility.* Successful teams are made up of team members who take responsibility for their own performance and that of the team. Ineffective teams invariably have members who undermine the team's performance by refusing to take responsibility and blaming others when things go wrong.
- *Cooperativeness.* People who work together must cooperate with each other. It is always easier to pull the load when all team members are pulling on the same end of the rope. Team members who refuse, either overtly or covertly, to cooperate with their teammates undermine the effectiveness of the team.
- *Initiative.* Initiative means recognizing what needs to be done and doing it without waiting to be told. Team members with initiative never say, "That's not my job."
- *Patience.* One of the most difficult challenges faced by team members is learning to work together on a daily basis. It is always easier not to get along than it is to get along. Getting along takes work, and *human-relations* work takes patience.
- *Resourcefulness.* Resourceful people are those who find ways to get the job done regardless of obstacles. Because resources such as time, talent, and funding are often in short supply, resourceful people are always welcome on teams.
- *Punctuality.* People who are punctual (on time, on schedule) show respect for their team members, their customers, and other stakeholders. A team cannot function effectively unless its members show up for work on time and complete their work on schedule.
- *Tolerance/sensitivity.* The modern workplace is a diverse environment, and people in teams can be different in a lot of ways (e.g., race, gender, religion, culture, age, politics, and so on). Diversity can strengthen a team provided that team members are tolerant of and sensitive to individual differences. Team members who can relate only to people like themselves do not make good team players.
- *Perseverance.* To persevere is to persist unrelentingly in completing a task, regardless of obstacles. This is an important trait because every team faces obstacles every day. Team members who persevere in spite of the obstacles inspire other members who might want to give up.

LEADING TEAMS

Leading a team is not about bossing; it's about coaching. Like effective coaches of athletic teams, effective team leaders focus on developing the capabilities of team members and continually improving their performance. They approach their job from the perspective of achieving peak performance on a consistent basis. This approach is translated into everyday behavior, as shown in Figure I.1. The various coaching duties are explained in the following paragraphs.

Clearly Defined Charter

One can imagine a baseball, basketball, soccer, or football coach calling his team together and saying, "Our mission this year is to win the championship." In just one simple statement, the coach has summarized the team's mission. The coach didn't say the team would improve its record by 50 percent or its league standing by two places. He said the team's mission was to win the championship. Now all the team's actions should serve to accomplish this mission. Any action that does not is inappropriate. Coaches of teams in the workplace should be just as specific when defining their team's mission. The mission is an important component of the team charter.

Team Development/Team Building

Athletes on sports teams, from high school through the professional ranks, practice constantly. Practice is the most consistent aspect of their lives. During practice, coaches work on developing the skills of individual team members and of the team as a whole. Team development and team-building

- Coaches give their teams a clearly defined charter.
- Coaches make team development and team building a constant activity.
- Coaches are mentors.
- Coaches promote mutual respect between themselves and team members and among team members.
- Coaches make human diversity within a team a plus by dealing with it openly and positively.

FIGURE I.1 Coaching duties.

activities are ongoing and perpetual. Coaches in the workplace should follow the lead of their counterparts from the world of sports. Developing the capabilities of individual team members and continually improving the performance of the team as a whole should be a regular part of the job.

Mentoring

Good coaches are mentors. This means they establish a caring, nurturing relationship with team members. Mentoring activities include developing team members' job-related skills; building character; helping team members see the big picture and where they fit into it; helping team members develop success-oriented attitudes; giving team members insight into corporate culture; and helping team members learn to get along with people who are different from themselves.

Mutual Respect

Team members do not have to like their team leader, but they must respect her. Correspondingly, the team leader must respect team members. Respect must be a two-way street. To build mutual respect, team leaders must appreciate team members as assets, communicate with them openly and frequently, and exhibit the highest ethical standards (and expect team members to follow suit).

Human Diversity

Human diversity can be a plus in teams, but it is seldom a plus automatically. People who differ in terms of culture, religion, gender, race, politics, and worldview often differ in their points of view about work-related issues as well. The natural human tendency is to gravitate toward people who look, think, and act like ourselves. Although there is comfort in uniformity, there is also danger. Teams perform best when their members bring diversity of thoughts, opinions, and ideas to the table. A diverse team is more likely to view problems from a variety of perspectives than is a uniform team, and different perspectives are important when attempting to solve problems and improve performance.

Team leaders should work to prevent the natural tendency of people to erect barriers between themselves and others who are different. This can be done by conducting a cultural audit to identify the demographics, personal characteristics, cultural values, and individual differences among team members; identifying the specific needs of individual team members; confronting cultural clashes openly, immediately, and forthrightly; and working to eliminate institutionalized bias.

THE TEN-STEP MODEL FOR EFFECTIVE TEAMWORK

To survive and thrive in a globally competitive marketplace, technology companies must perform consistently at peak levels—and their teams must do the same. What follows is a ten-step model for effective teamwork that can be used by technology companies and technical professionals.

1. Establish direction and goals for the team.
2. Establish clear roles and ground rules for the team.
3. Establish accountability for the performance of the team.
4. Develop team leadership skills.
5. Develop communication skills for team leaders and team members.
6. Develop conflict-resolution skills for team leaders and team members.
7. Establish a well-defined decision-making process for the team, and empower team members to be part of the process.
8. Establish positive team behaviors, ethics, and trust among team members.
9. Recognize and reward effective team performance.
10. Continually evaluate, improve, and build the team.

Endnotes

1. Peter R. Scholtes, *The Team Handbook* (Madison, WI: Joiner Associates, 1996), p. 6-1.
2. Institute for Corporate Competitiveness, *Final Report: Team Success Study* (Niceville, FL: 1995), pp. 2–4.

Establish Direction and Goals

[People], like nails, lose their usefulness when they lose direction and begin to bend.

—Walter Savage Landor

OBJECTIVES

- Demonstrate how to write a team charter.
- Explain how to write a team mission statement.
- Demonstrate how to write team goals.
- Explain how to write team assignments.
- Demonstrate how to write team schedules and deadlines.

Picture the following scenario. You have just been hired by ABC Company and assigned to the Adams Team (so named because the team leader is Jeffrey Adams). While in college, you held a variety of odd jobs, but this is your first real job in your field, and you want to make a good first impression. To this end, you want to ask the right questions of Jeffrey Adams during your orientation. But what should you ask? What would a person newly assigned to a team need to know?

First, you would want to know the team's purpose (mission). Why does the team exist? You would also want to know what the team is supposed to accomplish (goals). And you would want to know specifically what you and other team members are responsible for (assignments). Finally, you would want to know the due dates of your assignments and those of your fellow team members (schedule and deadlines).

These would all be legitimate concerns for a newly hired person, and any questions about them would be legitimate questions. This type of information is documented in a team charter. This chapter explains how to develop an effective team charter.

TEAM CHARTER

The team charter is a brief document that gives direction to individual team members and to the team as a whole. It should contain the following information:

- Mission statement
- Goals
- Assignments
- Schedule/deadlines

Figures 1.1 and 1.2 are examples of team charters. Figure 1.1 is a charter for a work team. Figure 1.2 is a charter for a process-improvement team. Notice that both team charters include the name of the team captain and give sufficient information to explain the big picture in terms of the team's direction and responsibilities. The team charter is not a day-to-day chronicle of the team's activities. Rather, it is a *big-picture* document that presents the team's overall direction in broadly stated terms. (There will be numerous specific tasks and duties associated with each assignment; the team captain will give interim deadlines for those specific tasks and duties.) Every team, regardless of its type, should be provided a team charter.

Who Develops the Team Charter?

It is important that all teams have a charter, but who should develop the charter? Higher management? The team leader? The team itself? The answer to this question is simple: It depends. In some situations, higher management should develop the team charter. In other situations, the team leader is given the task. In yet other situations, the team participates in the development of a charter, which is then submitted to higher management for review.

Team Charter

City Hall Parking Deck Project

Team Captain: Anthony Smith

Mission Statement

- The mission of the *City Hall Parking Deck Team* is to complete the structural design and planning for a multilevel parking deck for Melville City Hall.

Goals

- Complete design calculations for all structural components of the parking deck (prestressed concrete).
- Complete the shop drawings for fabrication of all prestressed concrete components of the parking deck.
- Complete the erection drawings for the parking deck.
- Complete the bill of material for the parking deck.

Assignments

- Design calculations: Smith
- Erection drawing package: Jones and Chang
- Fabrication drawing package: Rodriquez and Matsuoko
- Bill of material: Washington
- Checking drawings: Smith

Timeline

Assignment	Deadline
■ Design calculations	January 15
■ Erection drawing package	March 1
■ Fabrication drawing package	March 30
■ Bill of material	April 10
■ Checking drawings	April 20

FIGURE 1.1 Sample team charter.

Team Charter

Defect Minimization Team

Team Captain: Cindy Brittson

Mission Statement

MicroTech Company spends more than $300,000 annually on rework tasks in its electronic assembly department. The *Defect Minimization Team* is an ad hoc team with the following mission: *To identify strategies for minimizing the number of defects that occur in the electronic assembly department.*

Goals

- Identify the most frequently occurring defects.
- Prioritize the identified defects according to frequency of occurrence.
- Recommend strategies for eliminating the most frequently occurring defects.

Assignments

- Flowchart assembly process: Brittson (coordinates all stakeholder input)
- Develop Pareto charts: Jones, Ishakocus
- Develop cascaded Pareto charts: Maxim, Moth, Perez
- Develop improvement strategies: All members

Timeline

Assignment	Deadline
■ Flowchart assembly process	March 1
■ Develop Pareto charts	March 7
■ Develop cascaded Pareto charts	March 14
■ Develop improvement strategies	March 21

FIGURE 1.2 Sample team charter.

When deciding who should develop a team charter, there is only one absolute: Regardless of who writes the charter, it must be approved by higher management. There are several reasons for this, but perhaps the most important is that only higher management can commit the resources necessary to fulfill the team's mission. In addition, the approval of higher management prevents the proliferation of "rogue teams" (teams that go their own way irrespective of the company's larger mission and goals).

The following approaches are recommended for the development of team charters for the different types of teams. These are not hard-and-fast rules; remember that there is only one absolute when developing team charters.

- *Work teams.* Higher management should provide the team leader with either a mission statement or sufficient information to develop one. The team leader develops the goals, assignments, and schedule/deadlines (higher management provides the overall deadline for the project in question). The charter developed by the team leader is approved by higher management.
- *Improvement teams.* Higher management selects the team members and designates one of them as the team leader. The team leader is given guidance in the form of broad parameters and is then expected to draft a charter with the assistance of and input from all team members. The charter must be approved by higher management.
- *Standing committees.* Standing committees are usually chaired or staffed by a professional with the appropriate expertise. For example, the safety director would either chair or staff the company's safety committee. When this is the case, the related professional drafts the team charter for the approval of higher management. When there is no related professional for a given standing committee, higher management develops the team charter.

TEAM MISSION STATEMENT

The mission statement explains the team's reason for existing. A mission statement should be written in terms broad enough to encompass the team's total responsibilities, but specific enough to allow progress to be measured. Figure 1.3 lists the characteristics of a well-written team mission statement. The following sample mission statement for an improvement team has these characteristics:

> The purpose of this team is to reduce the time between when an order is taken and when it is shipped.

Characteristics of a Well-Written Team Mission Statement

- It answers the question: *Why does the team exist*?
- It is brief (no more than a paragraph).
- It explains *what* and *why*, not *how*.
- It is broadly stated. (Specifics show up in the team goals.)

FIGURE 1.3 A team mission statement should meet all these criteria.

TEAMWORK TIP *"If you can dream it, you can do it."*

—Walt Disney

TEAM GOALS

The details of the team's responsibilities are outlined in the team goals section of the charter. By informing team members of specifically what the team is responsible for, team goals—which support the mission statement—help members see how they fit into the *big picture* (something every employee wants and needs to know). Team goals should have the characteristics listed in Figure 1.4 and explained in the following paragraphs.

Characteristics of Well-Written Team Goals

- They are tied directly to the mission statement.
- They have a single-issue focus.
- They answer the following question: *What does the mission statement mean?*
- They show specifically what the team is supposed to accomplish.

FIGURE 1.4 Team goals should have these characteristics.

Team Goals Are Tied Directly to the Mission

Team goals should grow out of the mission statement so that they are tied directly to it. If a goal cannot be tied directly to the mission, either the goal is inappropriate or the mission statement is poorly written. To understand this point, consider the following team mission statement and partial set of team goals for a construction company:

TEAMWORK TIP

"Before you can score, you must first have a goal."

—Anonymous

MISSION STATEMENT

The *Customer Satisfaction Improvement Team* of Danville Construction Company is an ad hoc team with the following mission: To identify the root causes of customer satisfaction problems and propose strategies for eliminating the problems.

TEAM GOALS

1. Prioritize recorded customer complaints by frequency of occurrence.
2. Solicit customer feedback to identify any other complaints that have not been recorded.
3. Increase the company's rolling stock by 20 percent over the next two years.
4. Propose at least one strategy for correcting the five highest priority complaints.

The mission of this team is to find ways to improve customer satisfaction. The first two goals tie directly to this mission because they are actions that might reasonably be undertaken to find ways to improve customer satisfaction. The fourth goal is appropriate for the same reason. However, the third goal has no apparent tie to the mission.

The person drafting the team's goals might think that having more rolling stock would improve customer satisfaction, and in fact it might. But such an assertion has yet to be proven or even tested. If there is a tie between the mission and the third goal, it is too vague, to say the least. Consequently, the third goal should be either rewritten or dropped altogether.

Team Goals Have a Single-Issue Focus

Team goals, like all well-written goals, should focus on just one issue. Examine the following team goals:

- Determine the root cause of the solder-connection defects in all printed circuit boards in job lot 01-19C.
- Determine the root cause of the solder-connection defects in all printed circuit boards in job lot 01-19C, and identify a new connector mechanism for Type X-2 wire harnesses.

The first goal has a single-issue focus. The team is responsible for determining the root cause of defects in the solder connections in a specific batch of printed circuit boards. Progress made toward accomplishing this goal can be monitored, and the goal's eventual accomplishment can be measured.

The second goal, however, has a dual focus (solder-connection defects and a new connector mechanism). It should be broken up into two goals. When more than one major focus of action is contained in the same goal, both monitoring and measurement become more complicated.

Team Goals Explain What the Mission Statement Means

The team's mission statement is purposely broad. It shows team members the big picture, which is important to their understanding of the team's job. Having seen the big picture, team members then need more specific information. If you were taking an automobile trip from Florida to California, for example, to show your fellow travelers the big picture, you might open your road atlas to the map of the entire United States and the major east-west/north-south roadways. After they had seen the big picture, they would then want to look at the more detailed state maps, to understand more fully what the trip would entail.

Another way to understand the relationship between the level of detail contained in a mission statement compared with that contained in team goals is to consider the grocery store analogy. Assume your company is going to have an employee appreciation cookout. Several teams have been formed and given responsibility for various aspects of planning, implemention, and clean up for the event. You are on the Food Acquisition Team. The team's mission is to:

Go to the grocery store and purchase the food for the cookout.

This mission statement certainly provides the big picture, but it also raises a number of questions. At the very least, team members would want to know the types and quantities of food to purchase. These and other questions of a specific nature should be answered by the team's goals. Such goals might read as follows:

- Purchase 100 pounds of lean, packaged hamburger patties.
- Purchase 5 gallons of packaged potato salad.
- Purchase 5 gallons of packaged baked beans.
- Purchase 100 packages of hamburger buns.

Team Goals Are Specific Enough to Be Measured

Good team goals are written in a way that accommodates measurement. It should be easy to determine whether a team goal has been accomplished. For example, consider the goals from the previous section for the hypothetical Food Acquisition Team. The fourth goal read as follows:

Purchase 100 packages of hamburger buns.

Measuring the accomplishment of this goal requires nothing more than counting the number of packages of buns purchased. If the team has purchased 50 packages of buns so far, this goal is 50 percent accomplished. When the team has purchased 75 packages of buns, the goal is 75 percent accomplished, and so on, until the team finally locates and purchases a total of 100 packages of hamburger buns.

Measurement in this case is easy, because the goal was written to accommodate it. Just a minor change in the wording, however, could make this goal difficult or even impossible to measure. Consider the following reworded versions of the same goal:

- *Purchase hamburger buns for the company picnic.*
- *Purchase enough hamburger buns for the company picnic.*

In the first rewritten goal, there is no indication of quantity. Consequently, team members would have to ask, How many? Taken literally, this goal could be accomplished with the purchase of just one package of hamburger buns. When no amount is specified, any amount is right. In the second rewritten goal, the team is supposed to purchase "enough" hamburger buns for the company picnic, but how many will be enough? The only way to measure the accomplishment of the goal stated in this way would be to have the company picnic and see what happens. If the cooks do not run out of buns, the goal is accomplished (although the team might have purchased too many hamburger buns, resulting in costly waste). If the cooks run out of buns, the goal is not accomplished. This would be a form of measurement, but it would be an inefficient one at best.

Team Goals Show Specifically What the Team Is Supposed to Accomplish

Well-written team goals show team members what they—working together—are responsible for accomplishing. If a team's charter consisted only of assignments for individual members, the members would pursue their work as individuals. This is why the team's goals come before individual assignments in the charter.

First the big picture, then goals, then individual assignments. One of the ways goals promote teamwork is through *tactical mutual assistance*

(*TMA*). To understand this concept, consider the analogy of the basketball team. All five players on the court have their individual assignments. On offense, the point guard is responsible for bringing the ball up court and for setting up the plays. The center is responsible for offensive rebounds and driving the ball to the basket. If the point guard is double-teamed and has trouble getting the ball up the court, one of the other players on the team helps out temporarily. If the center has trouble rebounding, his teammates help him out. They do this because they all share the common mission of winning the game. If providing TMA is necessary to win, team members provide it. The same concept applies to work teams, improvement teams, and standing committees.

TEAM ASSIGNMENTS

The list of assignments is the most specific component of the team charter. There is an important balance to be achieved in developing the assignments. On the one hand, each team member must be able to see clearly what his typical assignment within the team is. On the other hand, assignments in a team charter are not so rigid that they cannot accommodate on-the-spot changes by the team's leader.

If the team in question were a baseball team, for example, each member would need to know who was assigned to play first, second, shortstop, and so on. A team charter would provide this information. However, these *typical* assignments would not mean that the coach might not move a player to another position for the good of the team. Of course, the coach may do so and, when necessary, would.

In addition, team assignments do not rule out the concept of TMA explained earlier. The assignment component of the team charter conveys information about the tasks that, on a given day unless otherwise specified by the team's leader, each employee should be working on. When the team leader gives a team member a temporary assignment other than the one designated in the charter, or when one team member steps out of her assignment to provide TMA to another member, the original assignment still applies: The team member returns to her typical assignment as soon as she completes her temporary duties.

Consider the example of an engineering team in a company that designs and erects prefabricated metal buildings. The team consists of one engineer (Hoyt), three CAD technicians (Andrews, Yamato, and Garcia), and a checker (Jones). For every building the company plans to construct, the team is responsible for design calculations and for developing a complete set of erection drawings. A typical set of assignments would be as follows:

- Design calculations: Hoyt (team leader)
- Foundation plan: Andrews

- Anchor-bolt plan: Andrews
- Rigid-frame sections: Yamato
- Roof-framing plan: Yamato
- Wall-framing plans: Yamato
- Connection details: Garcia
- Bill of materials: Garcia
- Checking plans: Jones

Making design calculations requires engineering expertise and credentials. Checking plans requires the type of knowledge that comes with experience. Consequently, only Hoyt and Jones would be given these assignments on this particular team. The assignments of Andrews, Yamato, and Garcia, however, are interchangeable. This means that although these three team members have specific assignments, there is excellent potential for TMA among them. It also means that the team leader can make temporary revisions to the usual assignments when necessary.

TEAM SCHEDULES AND DEADLINES

The degree of specificity of the schedules and deadlines in team charters depends on the type of team. With work teams, schedules and deadlines are usually tied to specific projects. Consequently, they must be updated when one project is completed and another is started. In the previous section, a typical set of assignments was presented for the development of a set of erection drawings for a prefabricated metal building. Those assignments are repeated below with corresponding deadlines.

Assignment	Deadline
Design calculations	January 15
Foundation plan	January 16
Anchor-bolt plan	January 17
Rigid-frame sections	January 20
Roof-framing plan	January 21
Wall-framing plans	January 22
Connection details	January 25
Bill of materials	January 27
Checking plans	January 30

The date that corresponds with each assignment represents a *hard stop* beyond which the work must not extend. This allows each individual to schedule his work to meet the deadline. It also allows other team members to see where each assignment should stand at a given point in time.

If the team member assigned the foundation and anchor-bolt plans completes them both on January 16, she can practice TMA and spend January 17 helping her teammates with their work. Because the team in this example is a work team, new deadlines will be established when the current project is completed and a new one is started. In fact, other deadlines might be assigned concurrently, because few work teams have the luxury of working on just one project at a time.

Schedules and deadlines are even more important for improvement teams and standing committees than they are for work teams, because such teams are ad hoc in nature. People who serve on ad hoc teams still have their *real* jobs to do. Consequently, they need to know precisely when their ad hoc work is due so that they can schedule it around their daily responsibilities. With ad hoc teams, it is important to schedule not just hard-stop deadlines, but intermediate checkpoints as well. This enables the team leader to monitor progress, because he typically does not interact with his ad hoc team members as frequently as work team leaders do.

Summary

1. A team charter gives a team direction. It contains a mission statement, goals, assignments, and a schedule or deadlines. A team charter ensures that team members see the big picture and where they fit into it.

2. A well-written team mission statement has the following characteristics: it explains why the team exists; it is brief (no more than a paragraph); it explains what and why, but not how; and it is broadly stated.

3. Well-written team goals have the following characteristics: they are tied directly to the mission; they have a single-issue focus; they explain what the mission statement means; they are specific enough to be measured; and they show specifically what the team is supposed to accomplish.

4. Team assignments are the most specific component of a team charter. They let individual team members know what their primary assignments are, but they do not rule out tactical mutual assistance (TMA). TMA occurs when team members lend each other a hand to get the team's work done right and on time.

5. Team schedules and deadlines are especially important for ad hoc teams because their members have their regular responsibilities to tend to in addition to their duties on the ad hoc team. Consequently, they need to know precisely when their ad hoc responsibilities must be completed. Work teams are typically given hard-stop deadlines that change as one project is completed and another is started.

Key Terms and Concepts

Big picture
Improvement teams
Single-issue focus
Specific enough to be measured
Standing committee
Tactical mutual assistance (TMA)
Team assignments
Team charter
Team goals
Team mission statement
Team schedules and deadlines
Tied directly to the mission
Work teams

Review Questions

1. What is the purpose of a team charter?
2. What are the necessary components of a team charter?
3. What are the characteristics of a well-written team mission statement?
4. What are the characteristics of well-written team goals?
5. What is the concept of tactical mutual assistance, and why it is an important aspect of teamwork?
6. What is a hard-stop deadline, and how is it used?
7. Why are schedules and deadlines even more important for ad hoc teams than for work teams?

EFFECTIVE TEAMWORK SIMULATION CASES

The following simulated cases deal with specific issues relating to the implementation of effective teamwork. Each case represents a meeting of Marcee McPhee and Pete Fared, engineers and team leaders at Mac-Tech, Inc., a technology firm with 526 employees. McPhee and Fared are not just colleagues; they are friends, and their friendship goes all the way back to college. They both attended the same engineering school and graduated in the same class. On the job, their relationship has evolved into one of mentoring, in which McPhee is helping Fared learn to be a better team leader. Once a week they meet for lunch and discuss problems, progress, issues, and concerns. These cases chronicle their luncheon conversations and invite the reader to discuss the issues Fared and McPhee deal with.

CASE 1.1 Team Members Need to See the Big Picture

"Marcee, how do you do it?" asked a frustrated Pete Fared. "Do what, Pete?" "How do you keep your team running so smoothly when you are out of the office? The minute I leave the building, my team seems to fall apart. Work slows down, team members get into turf battles, and quality goes out the window."

"Pete, let me ask you some questions." "Ask away," responded her colleague, anxious to pick up any tips McPhee might offer. "What is your team's mission?" "You know what our mission is, Marcee." "Yes, Pete, and so do you. But what about your team members? Do they know?" Pete Fared just shrugged. "What about the team's goals, Pete? Does every member of your team know what the team's goals are?" Once again Fared just shrugged. This discussion wasn't going anywhere, in his opinion. "What about the assignments and corresponding deadlines? Do all team members know their assignments and the timelines for completion?"

"Come on, Marcee. Are you going to help me or not?" snapped a frustrated Pete Fared. "Settle down and listen, Pete. I'm trying to show you why your team falls apart every time you leave the building." "All right, I'm sorry," mumbled Fared, contrite but still frustrated. "Look Marcee, I'm under a lot of pressure here. Every month my team's productivity and quality numbers are plotted on a Pareto chart and displayed with those of all the other teams. It's embarrassing when my team comes in dead last."

"Pete, my team can perform as well when I'm out of the office as when I'm in because every team member sees the big picture and knows where he fits into it. Every member also knows not just his own assignment but those of his teammates as well. This creates a kind of peer pressure within the team to perform well even when the boss is away. It also prevents turf battles and petty disputes over who should be doing what."

Discussion Questions

1. What types of problems can be anticipated if people serving on a team do not see the big picture?
2. Have you ever worked in a situation in which little got done when the boss was away? If so, did your unit, department, or team have a charter?
3. Have you ever worked in a setting in which you could not see the big picture? If so, did this cause any problems, concerns, or issues?

CASE 1.2 Writing a Team's Mission Statement

"Marcee, I knew I should have paid better attention when we were in college. Writing my team's mission statement is not as easy as I thought it would be," said a frustrated Pete Fared. He had been working on his mission statement for two hours, but was no closer to getting it right than when he had first started. McPhee recently gave Fared a copy of her team's charter to use as a guide in developing one of his own.

"The mission statement in your team's charter is brief and to the point, Marcee. But mine is a mess. It keeps growing. The longer I work on it, the longer it gets."

"I had the same problem when I was developing my team's mission statement. My first draft was one full page, typed," said McPhee.

"So, how did you reduce it from an essay to a paragraph?" asked Fared. "Well, I found my old classnotes from that course we took on quality management when we were in college. Our quality course had a whole unit on teamwork—part of which covered developing team charters. My notes contained an entry that I had underlined and circled. It said we should avoid inflating the mission statement with information that should be part of the goals or assignments," related McPhee.

"That must be where I'm going wrong," said Fared. "My mission statement has way too much detail in it." "Let me ask you a couple of questions that might help, Pete. What products are your team responsible for?" Fared told her. "Good," said McPhee. "Save that thought. Now, what processes do your team members perform?" "Same as yours," answered Fared. "That's right, and that's all you really need to know to write your mission statement, Pete. It's just that simple. The mission statement should tell why your team exists—which is to produce certain products or provide certain services—and state that it does this using certain processes. Put that information into the form of a mission statement and you'll have a good one," suggested McPhee.

Discussion Questions

1. Do you think it is important that the people serving on a team understand the team's mission?
2. What types of problems might be expected when team members do not understand the team's mission?
3. Why do you think it is important for a mission statement to be relatively broad, short, and simply written?

CASE 1.3 Fared Writes His Team's Goals

"Marcee, I don't know if the classnotes you loaned me are helping or just complicating things," said Pete Fared with a sigh. "Why? What's the problem?" asked McPhee. "It's all of these requirements," responded Fared, pointing to the notes.

"Tell me what's causing problems with your team's goals, Pete." "It's this checklist in your class notes: *Characteristics of Well-Written Team Goals.* Team goals are supposed to have a single-issue focus, among other things," responded Fared. "I'm OK with all of the other characteristics except this one. I'm not sure what it means." "Have you written any of your goals yet?" asked McPhee. "Just one," answered Fared. "I'm hung up on the first one, but if I can get it straightened out, I'll know how to do the others."

"Let's take a look at it," offered McPhee. Fared placed a sheet of paper on the table.

Design and plan a minimum of 60 XYZ units per month.

"I don't know if this goal has a single-issue focus or not," admitted Fared. "I mean, is the single issue the XYZ units, or is it the designing and planning?" "That's a good question, Pete. Let me answer it with another question. Can you envision a month in which the design element of the goal gets done, but the planning doesn't?" "Not only can I envision it, Marcee. It happens more times than I care to admit."

"Then the goal does not yet have a single-issue focus," said McPhee. "Designing XYZ units is one task and developing the plans that document that design is another task. You need to have a separate goal for each of these tasks, Pete. If not, you won't be able to measure your team's performance."

Discussion Questions

1. What problems might be expected in a team if the team's goals do not correspond closely with the team's mission statement?
2. What problems might be expected if a team's goals are not measurable?
3. Defend or refute the following statement: "If we have a good mission statement, we don't need team goals."

CASE 1.4 Work Assignments Can Backfire

"Marcee, do you remember when we first talked about developing a team charter?" "I remember, Pete." "One of my biggest problems was turf battles," said Pete Fared. "Whenever I was out of the office, my team members always seemed to get into these 'it's-not-my-job' squabbles." "Well, Pete, the assignments component of your team charter should have solved that problem." "I thought it would have too," said Fared with a sigh of exasperation. "But it didn't. In fact, now that I've written down who is supposed to do what, nobody wants to help anyone else. It's almost as bad as it was before I wrote down the assignments."

"Pete, I apologize. There is another aspect of making assignments I should have discussed with you, and it gets right to the heart of the problem you're having." Marcee McPhee went on to explain the concept of *tactical mutual assistance (TMA)* and how she instilled it in the thinking and everyday practice of her team members. She also explained that she had worked with the Human Resources Department to rewrite the job descriptions of her team members to include an expectation of TMA, adding it to the list of criteria she would address in performance appraisals. "Pete, TMA is one of the things that makes a team a team instead of just a collection of individuals."

"So, Marcee, you actually rewrote their job descriptions and changed the way you evaluate their performance?" "That's right, Pete. And I explained all of this to them as a group. No surprises, Pete. That's important." "How did your team members take it?" inquired Fared. "Well, there was some grumbling at first. Change is hard on people, even when it's change for the better," counseled McPhee. "But within a couple weeks, TMA was happening. Now I'm working with the Human Resources Department to revise the company's performance-appraisal form to include formal criteria on TMA and other teamwork-related issues, and to implement team-based financial incentives."

Discussion Questions

1. What problems might be anticipated when establishing tactical mutual assistance as a common practice in a team?
2. What reasons might team members give for not wanting to practice TMA?
3. What strategies might team leaders use to overcome resistance to TMA?

2

Establish Clear Roles and Ground Rules for Teams

In order for people to work well in teams, they must understand their roles and the ground rules. Ambiguity is the enemy of teamwork.

OBJECTIVES

- Explain the role of the team leader.
- Explain the roles of team members.
- Demonstrate how to develop ground rules for team members.

Few people naturally work well in teams. Effective teamwork requires many behaviors that can run counter to human nature, and it requires that these behaviors be applied on a consistent basis. Effective team behaviors include selflessness, trust, perseverance, cooperation, tolerance, and sensitivity. Most people can learn to work effectively in teams provided all the prerequisites are in place. Two of the most critical prerequisites are roles and ground rules. To work effectively in teams, people need to understand their roles, the role of their team leader, the roles of other team members, and the ground rules for the team.

THE TEAM LEADER

Team leaders can go by many different names, depending on the nature and preferences of the company, and on the type and size of the team. They may be called team leader or team captain; or they may be called supervisor, department head, division director, superintendent, project manager, unit head, committee chair, or even CEO. Regardless of the individual's title, every team must have a designated leader. This does not mean that individual team members do not apply leadership skills—quite the contrary. Team members can and should use whatever leadership skills they have to continually improve the team's performance. For the sake of role clarity and accountability, however, someone must be in charge.

Consider the example of a football team. The offensive and defensive teams each have their own designated leader. On offense, this individual is the quarterback. He calls the plays, communicates with higher management (the coach) on behalf of the team, adjusts the assignments of players when necessary, and coordinates all offensive action. On defense, usually the middle linebacker is the designated leader.

Every other player on the team is encouraged to use whatever leadership skills he has to help the designated leader (the quarterback or the middle linebacker) keep team members motivated, encourage other players to persevere when the going gets tough, and assist teammates who are struggling with their assignments. On the other hand, no team member, no matter how good a leader, should do anything to undermine the authority or effectiveness of the designated team leader. Team members should seek to supplement, not supplant, the leadership of the team leader.

TEAMWORK TIP

"Managers are people who do things right, and leaders are people who do the right thing."

—Warren G. Bennis and Burt Nanus, leadership experts

Role of the Team Leader

Team leaders play a dual role: one role as a team member performing her daily work duties, and the other role as the leader. The better they perform the former, the better they will be able to perform the latter. To understand the dual nature of the team leader's role, consider again the example of the football team. The quarterback has his normal duties, involving, among other things, passing, handing off, and running the football. In addition, he is expected to provide leadership for the team. In this role he attempts to inspire his teammates by setting a good example, motivate players to perform consistently at peak levels, observe the performance of individual players to determine if they need help immediately or extra training later, consult with higher management (the coach) on tactical issues (what play to run in a given situation), and offer input to higher management on strategic issues (formulating a game plan).

Peter Scholtes describes the role of the team leader as follows:

> The team leader is the person who manages the team: calling and facilitating meetings, handling or assigning administrative details, orchestrating all team activities, and overseeing preparations for reports and presentations. The team leader should be interested in solving the problems that prompted their project and be reasonably good at working with individuals and groups. Ultimately it is the leader's responsibility to create and maintain channels that enable team members to do their work.[1]

TEAMWORK TIP

"Authority without wisdom is like a heavy ax without an edge, fitter to bruise than polish."

—Anne Bradstreet, American author

Although in this passage Scholtes is writing about a temporary or ad hoc team (such as an improvement team), his description applies equally to work teams. The team leader of any kind of team also has the following responsibilities:

- Setting a positive example for team members
- Serving as liaison between the team and the rest of the organization
- Maintaining the team's records
- Monitoring the performance of the team as a whole
- Monitoring the performance of individual team members
- Making assignments
- Providing a charter that gives direction and the big picture
- Evaluating the performance of team members
- Arranging training for the team and for individual members
- Keeping the team's work on schedule
- Keeping the team informed concerning issues within the company
- Arranging mentoring relationships within the team
- Managing conflict within the team
- Building trust among team members

- Establishing and maintaining a positive work environment for the team
- Encouraging and reinforcing positive team behaviors
- Ensuring appropriate recognition for the team and its members
- Taking responsibility for the team's performance (accountability)

TEAM MEMBERS

It is important that team members make a positive contribution to the team. The following strategies help individuals in the workplace participate effectively on a team:[2]

- *Gain entry.* Get acquainted with your fellow team members as soon as possible. Let them know who you are and what you can contribute, but more importantly, find out who *they* are and what *they* can contribute.
- *Be sure you have a clear understanding of the team's mission.* Team members cannot contribute to the team's mission if they don't know what it is. Learn the mission statement; know the goals; understand assignments and timeframes; and communicate problems, progress, and other important information freely.
- *Be well prepared, and participate.* Good team members never wing it at team meetings. Before attending a meeting, prepare. Familiarize yourself with the agenda, read through the minutes of the last meeting, review any personal notes you might have taken, and write down any concerns you have or issues you want to raise. During the meeting, state what you have to say accurately and succinctly.
- *Stay in touch.* Good team members stay in touch between meetings and communicate frequently. Keep fellow team members up to date on your progress, and ask for their help with problems.

TEAM GROUND RULES: THE RATIONALE

There is an old saying, "Good fences make good neighbors." One of the reasons good fences make good neighbors is that fences establish clear boundaries. This is important because people like to know where their boundaries are and what they can do and say without stepping on someone else's toes or wandering onto someone else's turf. Boundaries are especially important when people interact in groups.

Group interaction magnifies the opportunities for misunderstanding, miscommunication, discord, strife, and conflict. This is because every individual in a group has her own personality, goals, motivations, ambitions, and way of doing things. Add to this the fact that people differ in so many ways (e.g., gender, race, religion, culture, national origin, politics, and so on), and you begin to see why clear boundaries are so important in groups.

A team is a group of people with a common mission. Boundaries are important in accomplishing that mission. In teams, boundaries are established by the adoption of a set of ground rules. There are several possible approaches to developing ground rules for teams. Some companies allow individual teams to develop their own ground rules. This approach has the benefit of *buy-in:* when teams develop their own ground rules, they are more likely to accept them and to enforce them through peer pressure. On the other hand, this approach lacks the advantage of uniformity. When every team has a different set of ground rules, problems among teams can occur, especially when individuals serve on more than one team. In an attempt to prevent such problems, some companies develop uniform guidelines that are distributed to all teams. This approach has the benefit of companywide uniformity, but it can fall short when it comes to employee buy-in. Employees tend to view company-imposed ground rules as "your rules, not mine."

These two approaches seem to be mutually exclusive and to represent an irreconcilable dilemma. This is not necessarily the case, however; the author recommends a third approach—one that, if used properly, can achieve both employee buy-in and companywide uniformity. This approach is explained in the next section.

DEVELOPING TEAM GROUND RULES

Ground rules do little good unless team members believe in and accept them. Consider the law on speed limits. How many people obey the rules when it comes to posted speed limits? Not many. Of course, the number is higher when a police officer is visibly present, but how about the rest of the time—typically the majority of the time? Employees who accept ground rules obey them even when an authority figure is not watching. This is the concept mentioned earlier, buy-in, which argues for letting each individual team develop its own ground rules.

On the other hand, a certain amount of companywide uniformity is desirable in team guidelines. If every team has its own guidelines and they differ radically, the differences can create problems. This is particularly true when one team is perceived as being strict with its members and another team is perceived as being lax. It is desirable that team ground rules promote both buy-in and uniformity.

Fortunately, this mix can be achieved, at least to an acceptable extent. To develop ground rules that achieve both buy-in and an acceptable measure of uniformity, companies can apply the guidelines shown in Figure 2.1. These guidelines are explained in the following paragraphs.

Guidelines for Developing Team Ground Rules

- Form a cross-functional ad hoc committee to develop a standard list of issues teams should consider when developing ground rules.
- Circulate the draft list of issues among all employees companywide, and ask for their input.
- Have the ad hoc committee finalize the standard list of issues based on employee input.
- Give all team leaders the standard list of issues to use as a guide when working with their respective teams to develop ground rules.

FIGURE 2.1 The recommended method for developing team ground rules.

Form a Cross-Functional Ad Hoc Team

A cross-functional committee or team is one comprised of at least one representative from each functional unit in a company with a stake in a given issue. Cross-functional groups are useful when it is important to ensure broad-based input and representation (this promotes buy-in while simultaneously improving the quality of the input). In the present case, the cross-functional ad hoc committee is responsible for developing a standard list of issues that teams should consider when establishing their ground rules. Like any team, this ad hoc committee should be given a written charter. Because all functional units in the company are represented on the committee, the company's CEO should develop the charter. If the team members represented a division, branch plant, or some other unit within the company, the responsible manager would develop the charter. Figure 2.2 is an example of such a charter.

Circulate the Draft List of Issues

Once the Ground Rules–Issues Committee has a draft list of issues that is acceptable to its members, the list is circulated among all employees company- or unitwide. Recipients of the list are asked to add or eliminate issues, and to indicate any other input they deem pertinent. All of this is done without attribution, to encourage candor. Employees are given a

Team Charter
Ground Rules–Issues Committee

Committee Chair: Victoria Hansen

Mission

The mission of the *Ground Rules–Issues Committee* is to develop a standard list of issues teams should consider when establishing their ground rules.

Goals

1. Develop a draft list of ground rules issues.
2. Circulate the draft list of ground rules issues among all employees for their input.
3. Finalize the list of ground rules issues based on employee input.

Assignments

All members of the committee will participate in all activities of the committee.

Schedule

1. Draft list . . . . January 15
2. Circulation of draft list . . . . January 30
3. Final list . . . . February 15

FIGURE 2.2 An example of a team charter.

deadline for this task. Figure 2.3 is an example of a list developed by a ground rules–issues committee.

Finalize the List of Issues

Once input from employees has been collected, the committee meets to consider suggested additions and deletions. Those that are ratified by the committee are used to produce a final list. The final list is presented to the company's CEO or the responsible unit manager by the committee's chair, along with any pertinent input employees might have submitted in the form of written comments.

Issues to Consider for Team Ground Rules

- Honesty
- Selflessness
- Unity
- Cooperation
- Diversity
- Punctuality
- Perseverance
- Conflict resolution
- Attendance
- Respect
- Leadership
- Communication
- Trust
- Dependability
- Responsibility
- Initiative
- Resourcefulness
- Tolerance/sensitivity
- Supportiveness
- Awareness
- Participation
- Courtesy
- Followership

FIGURE 2.3 These factors might be included in a team's ground rules.

Give the Final List to Team Leaders

The final list of ground rules issues is given to all team leaders with instructions to use the list as a guide in developing team ground rules. *Power of suggestion* typically ensures that all or at least most, of the issues on the list will appear in the ground rules statements developed by individual teams. On the other hand, because each team develops its own ground rules (using the list of issues as a guide), buy-in is virtually assured.

The final ground rules statements developed by individual teams will not necessarily be—and need not be—exact replicas of each other. There is more than one way to say the same thing. One team might treat each issue from the list separately, whereas another team might combine two or more of the issues. Figures 2.4 and 2.5 are examples of partial ground rules statements that at first glance look quite different but in reality say essentially the same thing.

Ground Rules Statement
Safety Committee

Honesty/Trust

Members of this team will be honest with each other so as to build trust.

Dependability/Punctuality

Members of this team will arrive on time every time and conduct themselves dependably in completing all tasks, responsibilities, and assignments.

Respect/Tolerance/Sensitivity

Members of this team will treat each other with respect, tolerance, and sensitivity in all aspects of their day-to-day interaction.

Unity/Support/Cooperation

Members of this team will promote unity through mutual support and cooperation in getting the job done.

FIGURE 2.4 Partial ground rules statement (first example).

Ground Rules Statement
Design/Planning Team

Trust

Team members will build trust through honesty, dependability, cooperation, and mutual respect.

Responsibility

Team members will take responsibility for their work and that of the team by displaying selflessness, good attendance/punctuality, active participation, and mutual support.

Tolerance

Team members will maintain an environment of tolerance by being sensitive to and respectful of individual differences.

FIGURE 2.5 Partial ground rules statement (second example).

Summary

1. Team leaders go by many different names, including team captain, team leader, supervisor, department head, division director, superintendent, project manager, unit head, committee chair, and so on. Regardless of what the individual is called, every team must have a leader. Team members, even though they are not the designated team leader, should apply whatever leadership skills they have to supplement, but never supplant, those of the team leader.

2. Team leaders play a dual role. One role consists of their day-to-day duties on the team. The other role is that of the team leader. The better an individual performs the former, the more credibility she will have in performing the latter.

3. Ultimately, the team leader is responsible for ensuring the consistent peak performance of the team and for continually improving that performance. Specific duties of a team leader include the following: setting a positive example, serving as liaison between the team and the rest of the organization, maintaining the team's records, monitoring the performance of the team as a whole, monitoring the performance of individual team members, providing direction and making assignments, evaluating the performance of team members, arranging for training, keeping the team's work on schedule, keeping team members informed, arranging for mentoring, managing conflict, building trust among team members, establishing and maintaining a positive work environment, reinforcing positive team behaviors, ensuring appropriate recognition, and taking responsibility for the team's performance (accountability).

4. The following strategies help individuals be effective team members: gain entry, be clear on the team's mission, be well prepared, participate, and stay in touch.

5. The rationale for having ground rules for teams is that people need to know what the boundaries are. They need to know what is expected of them and their fellow team members, and they need to know that the expectations are the same for every person on the team. When establishing ground rules, it is important to achieve both uniformity and buy-in. This can be done by giving team leaders a standard set of issues to use as a guide in developing their team ground rules and then allowing each team to develop its ground rules based on the guide.

Key Terms and Concepts

Be clear on the team's mission

Be well-prepared and participate

Buy-in

Circulate the draft list of issues

Finalize the list of issues
Form a cross-functional ad hoc team
Gain entry
Give the final list to team leaders
Good fences make good neighbors
Ground rules–issues committee
Role of the team leader
Stay in touch
Team ground rules
Team leader
Uniformity

Review Questions

1. What is the role of the team leader?
2. List ten specific duties of a team leader.
3. What strategies can an individual apply to become an effective member of a team?
4. What is meant by the term "buy-in" as it relates to ground rules?
5. What is meant by the term "uniformity" as it relates to ground rules?
6. How can a company achieve both buy-in and uniformity when establishing team ground rules?

EFFECTIVE TEAMWORK SIMULATION CASES

The following cases deal with specific issues relating to the implementation of effective teamwork. Each case represents a meeting of Marcee McPhee and Pete Fared, engineers and team leaders at Mac-Tech, Inc., a firm with 526 employees. McPhee and Fared are not just colleagues; they are friends, and their friendship goes all the way back to college. They both attended the same engineering school and graduated in the same class. On the job, their relationship has evolved into one of mentoring, in which McPhee is helping Fared learn to be a better team leader. Once a week they meet for lunch and discuss problems, progress, issues, and concerns. These cases chronicle their luncheon conversations and invite the reader to discuss the issues Fared and McPhee deal with.

CASE 2.1 Coaching versus Bossing a Team

"Marcee, do you ever have problems with your team members wanting to know why or asking a thousand questions every time you tell them to do something?" "What do you mean, Pete? Can you be more explicit?" asked Marcee McPhee. "Well, I tell my team members to do this or do that and they immediately start asking questions. They always want to know why." "And this bothers you, Pete—you think it's odd?" "Yes it bothers me, Marcee. I think when the boss says to do something, employees should snap to and do it. In fact, let me tell you a story," said Pete Fared. "I was in the Navy before you and I met in college. I joined the Navy to get the GI Bill to pay for college. Anyway, I worked for a Captain who expected us to jump when he said jump. We did what we were told, when we were told, and we didn't ask a lot of questions."

"Did this Captain ever ask for your input or opinion Pete?" "You've got to be kidding, Marcee. I once tried to give some input, and the Captain told me if he wanted me to have an opinion he'd give me one." "Well, I don't know anything about the military, Pete, but I do know that our company is not the Navy. For one thing, our employees can quit, and if we treat them like this Captain treated you, they will." When Fared had no response, McPhee asked, "Could you quit when you were in the Navy?" "No way. Once you signed the enlistment papers, the Navy owned you for four years. And I'll tell you something else, Marcee. For those four years, if I didn't jump when an officer said to, he could have me pushing a mop every weekend for a month."

"When you were in the Navy, did you like being ordered around without being asked your opinion, without being given the opportunity for input, or without being told why?" "Of course not!" said Fared. "In fact, I hated it." Once again McPhee could tell she had touched a nerve. "Now don't get me wrong, Marcee. I don't want to give you the impression that every officer in the Navy was some kind of tyrant. The one I worked for was, but he might have been the exception to the rule. I don't know. But I do know this: there must have been at least a hundred times when I could have gotten the job done faster, easier, and better if someone in charge had bothered to ask for my input. In fact, we had a name for officers like my Captain," said Fared with a smirk. "We called them 'seagull officers.'" Seeing the quizzical look on McPhee's face, her colleague explained. "A seagull officer is one who flies in where he isn't needed, makes a lot of noise, messes all over everything, and flies away." McPhee laughed out loud at the analogy and told Fared that she had worked for a few seagulls herself. Then she asked, "Would you like to work for a seagull manager here at Mac-Tech?" "Not on your life, Marcee." "Well, nor would your team members, Pete."

Discussion Questions

1. Would you rather be bossed or coached? Why?
2. Have you ever been in a situation where you were given orders but never asked for your opinion or input? If so, how did this approach affect your desire to get the job done?
3. Which approach do you think will produce the best results in a team in the long run, bossing or coaching?

CASE 2.2 Ground Rules Can Cause Problems

Pete Fared was in a foul mood when he joined his colleague, Marcee McPhee, for their weekly lunch meeting. "Marcee, we have a problem," began Fared without preamble. Holding up both hands to cut him off, McPhee responded, "I know, I know. I've already heard." Fared saw immediately that his colleague was frustrated too. Earlier McPhee had given Fared a copy of the ground rules her team had developed. He, in turn, had used them as a guide in helping his team develop ground rules of its own. "Marcee, I probably should have had my team just adopt your ground rules as written. That would have saved us both a lot of problems."

Actually, the team meeting Fared had called to develop ground rules had gone well. His team members had quickly warmed to the idea and had really gotten into the process. The meeting allowed several members to voice minor resentments they had been harboring about such issues as attendance, tardiness, and providing tactical mutual assistance when appropriate. Team members were also pleased to realize that henceforth they would all be subject to the same ground rules. All in all, Fared had been quite pleased with the results of the meeting and the process. Consequently, he was shocked to learn, just two days later, that his team was feuding with McPhee's team over the issue of ground rules.

Ironically, the tension between the two teams had gotten its start over a grilled chicken salad. A member of Fared's team is married to a member of McPhee's. Over supper on the day Fared's team had developed its ground rules, this husband and wife were having a what-happened-at-work-today discussion. The conversation eventually got around to the subject of ground rules, and the couple began comparing notes. What started as a simple discussion eventually escalated into an argument over why one team had certain ground rules and the other didn't. The debate then escalated even further into a verbal battle over which team's ground rules were better. At work the next morning, what had started out as a simple marital squabble quickly spread through both teams and became a workplace feud. By lunch, Fared and McPhee were dealing with a major distraction.

Discussion Questions

1. Would you be more likely to accept and support team ground rules you helped develop or ground rules that were simply given to you? Explain your reasoning.
2. Why do you think employees might object when one team's ground rules are noticeably different from another's? What problems might arise?
3. What are some of the important issues you think should be covered by a team's ground rules?

Endnotes

1. Peter R. Scholtes, *The Team Handbook* (Madison, WI: Joiner Associates, 1996), pp. 3–8.
2. Mary Walsh Massop, "Total Teamwork: How To Be a Member" in *Management for the 90s: A Special Report* from Supervisory Management (Saranac Lake, NY: American Management Association, 1991), p. 8.

3

Establish Accountability

One person seeking glory doesn't accomplish much. Success is the result of people pulling together to meet common goals.

—John C. Maxwell

OBJECTIVES

- Explain how to build a foundation for accountability.
- Distinguish between formal and informal accountability.
- Explain the concept of accountability for team leaders.
- Explain the concept of accountability for team members.
- Explain the concept of peer accountability.

The team charter and team ground rules establish what is expected of the team as a whole and of its members in particular. The team leader is expected to ensure that the team fulfills its charter properly and on time. Individual team members are expected to do their work and conduct themselves in ways that support fulfillment of the team's charter.

When there are expectations, there must be *accountability*. Expectations without accountability are like a tiger without teeth. Accountability is a key element in the formula for successful teamwork. Team leaders, the team as a whole, and individual team members must be held accountable for the performance of the team. Accountability, by its very nature, implies consequences—either positive or negative based on how individual behavioral decisions are made and how actions taken affect the performance of the team. This chapter explains how to establish a comprehensive system of accountability for teams.

BUILDING A FOUNDATION FOR ACCOUNTABILITY

One of the worst mistakes a company can make is to hold teams accountable for performance without giving them the support and resources needed to perform. An effective program of team accountability is like a house: it must be built on a solid foundation. A team has a solid foundation for accountability when it fulfills the requirements listed in Figure 3.1.

The four components in Figure 3.1 form the necessary foundation for holding teams accountable for their performance. The team charter and team ground rules represent the first two elements of the foundation. The team's mission, as stated in its charter, explains what is expected of the team. The team's ground rules explain the behavioral parameters within which the mission will be pursued, and the team's goals show clearly how successful fulfillment of the mission will be determined (i.e., all goals accomplished properly and on time).

Training in both teamwork and normal work-related skills is also a critical element of the foundation for accountability. Team members cannot do what they don't know how to do. Chapters 4 through 8 of this book deal in part with various aspects of training for team members. Resources make up the final component of the foundation for team accountability. Well-maintained facilities, up-to-date equipment, proper tools, and sufficient time are necessary for successful performance. Teams cannot simply requisition these things; resources are provided by higher management.

Elements that Make Up the Foundation for Accountability

- Team members understand what is expected of them individually and of the team as a whole.
- Team members understand how fulfillment of expectations will be determined.
- Team members have been given the training necessary to fulfill the expectations.
- The team is given the resources and support necessary to fulfill the expectations.

FIGURE 3.1 An effective accountability system requires a solid foundation.

TEAMWORK TIP

Four of the key determinants in successful teamwork are the following:

- Personnel determine the potential of the team.
- Vision determines the direction of the team.
- Work ethic determines the preparation of the team.
- Leadership determines the success of the team.

—John Maxwell, teamwork expert

FORMAL VERSUS INFORMAL ACCOUNTABILITY

There are both formal and informal types of accountability. Formal accountability involves two processes: (1) written evaluations of the team's performance based on accomplishment of the team's charter and (2) evaluations of team members based on adherence to the team's ground rules. Ad hoc team leaders receive formal evaluations from the responsible higher manager once their teams have accomplished what they were chartered to do. Work-team leaders are evaluated by their supervisors as part of the regular performance-appraisal process. Members of ad hoc teams are evaluated by their team leader at the conclusion of their assignment. This evaluation is then considered when the team members are evaluated by their supervisors as part of the regular performance-appraisal process.

Formal evaluations are an important part of team accountability. There is a wise old saying: "If you want to lose weight, you have to step on a scale." This practical philosophy also applies to accountability. If you want to ensure and improve performance, you must measure it. There is no accountability without measurement, and formal evaluations involving specific criteria constitute that measurement.

In addition to formal evaluations, it is important to establish informal accountability in teams. Informal accountability has two components: (1) ongoing monitoring by the team leader, and (2) ongoing monitoring by fellow team members (peer pressure).

Ongoing Monitoring by the Team Leader

Formal, periodic, written evaluations are an important part of team accountability. However, written evaluations are typically done after-the-fact. They serve an important purpose, but they do not serve to improve performance until the next team assignment, if at all. To improve performance on the spot and in real time, team leaders should monitor the work of their team members as it is happening and provide verbal feedback—both positive and corrective—right then. This is especially true for work teams.

With ad hoc teams and standing committees, team leaders can monitor performance by asking for face-to-face progress reports at all team meetings and email or telephone updates between meetings. The key is to provide both encouragement and correction on the spot, so that performance improves immediately.

Ongoing Peer Monitoring

Teams work best when team members enforce their own rules through peer monitoring. People are susceptible to peer pressure for most of their lives. This is as true of positive peer pressure as it is of negative. The root of effective peer monitoring is strong, close, caring relationships among team members. Once team members care about each other, they will care about what their colleagues think of both their work and their behavior. On the other hand, trying to use peer monitoring in a team of discordant individuals who neither trust nor care about each other usually escalates the bad feelings that already exist.

Team leaders must judge when an appropriate level of trust has developed among their team members. Having judged that sufficient trust exists, team leaders can tell team members that they are expected to help each other perform at peak levels on a consistent basis by: (1) setting a good example for each other and (2) expecting every person on the team to live up to the ground rules and to carry his share of the load.

ACCOUNTABILITY FOR TEAM LEADERS

Provided the proper foundation for accountability has been established, the team leader is the person with primary responsibility for the team's performance. Along with responsibility comes accountability—two sides of the same coin. Team leaders are like football, baseball, and basketball coaches in that they get the blame when their teams fail to perform, even when it's really not their fault.

In the workplace, accountability for a team's performance should be based on a formal assessment of the team's actual performance compared with its expected performance as set forth in the charter. This comparative assessment is the first component in the accountability equation for team leaders. The second component is the team leader's regular, periodic performance appraisal.

Accountability Based On Team Performance

Measuring a team's performance is a matter of assessing the extent to which the goals set forth in its charter have been accomplished. This is the case regardless of the type of team in question—work team, ad hoc improvement team, or standing committee. Figure 3.2 contains goals excerpted from a team charter for a work team. The actual charter for this team contains more goals than are shown in Figure 3.2. These particular goals are excerpts for

Goals Excerpted from a Team Charter

- Complete the design calculations for Project 02-XT by January 15.
- Complete the drawings for Project 02-XT by February 10.
- Complete the bill of material for Project 02-XT by February 17.

FIGURE 3.2 Teams should have specific measurable goals.

the purpose of illustrating the role team goals play when assessing team performance for the purpose of accountability.

Examine the goals for the work team in Figure 3.2. The team is responsible for the engineering aspects of a construction project, Project 02-XT. The team is to complete the design calculations, drawing package, and bill of material, each by a specified date. In the case of the design calculations, this means the calculations must be performed, checked, corrected, and finalized no later than the specified date. The drawing package and bill of material must also be prepared, checked, corrected, and finalized by their specified dates. Meeting these deadlines is critical for two reasons. First, the work of one set of team members is dependent on the work of other team members. For example, the CAD technicians cannot complete the drawing package until they have the finalized design calculations. Correspondingly, the individuals assigned the bill of material cannot complete their work until they have the completed drawing package. Second, the work of other teams that purchase the materials, schedule the builders, and construct the building depends on the completed work of the engineering team. Consequently, the leader of the work team in this example is responsible (and accountable) for ensuring that his team completes its tasks not just properly, but also on time.

The manager who supervises the leader of the work team in this example evaluates the team leader by comparing actual performance against expected performance from the team's initial charter and all subsequent charters given the team. Were the design calculations for Project 02-XT completed by January 15? Were the calculations correct? If the answer to these questions is "yes," the team accomplished its first goal and the team leader should receive the appropriate credit. If the answer is "no," the responsible manager notes the failure and holds the team leader accountable.

Of course, evaluations of team performance are like all other performance evaluations in that there is room for some negotiating. When uncontrollable factors intervene and prevent or inhibit accomplishment of a goal, the team leader is responsible for immediately apprising his supervisor of the circumstances and negotiating appropriate revisions to the charter. This step is critical because the team leader is held accountable for the expectations set forth in the charter. The only way to change those expectations is to change the charter. Until that is done, the original expectations apply. This is as true of ad hoc improvement teams and standing committees as it is of work teams.

TEAMWORK TIP *"One accurate measurement is worth a thousand expert opinions."*

—Grace Murray Hopper, Admiral, U.S. Navy

The Team Leader's Periodic Performance Appraisal

Team leaders typically have other responsibilities beyond their team-related duties. This is particularly true for leaders of ad hoc improvement teams and standing committees. Consequently, team leaders undergo periodic performance appraisals that address all aspects of their jobs. It is important to include team leadership as part of these periodic performance appraisals, even if it means revising the company's performance-appraisal forms. Figure 3.3 is an example of a criterion that could be included in the performance-appraisal form for employees who serve as leaders of any type of team in the workplace. There are many other ways such a criterion might be stated. The key is that team leadership is represented by at least one criterion in the regular performance-appraisal process for team leaders.

Team Leadership					
3	6	9	12	15	*Rating*
Usually fails to meet expectations	Often fails to meet expectations	Occasionaly meets expectations	Often meets expectations	Consistently meets expectations	

FIGURE 3.3 Sample performance-appraisal criterion for team leaders.

ACCOUNTABILITY FOR TEAM MEMBERS

Whereas accountability for team leaders begins with the charter, accountability for team members begins with the ground rules. The ground rules are as important for assessing the performance of team members as the charter is for assessing the performance of the team leader. This is yet another reason why it is so important to develop a comprehensive set of team ground rules and to involve all team members in the development of those ground rules. Figure 3.4 is a list of criteria, based on a hypothetical set of team

Performance Appraisal Criteria for Team Members

5 = Always 4 = Usually 3 = Sometimes 2 = Seldom 1 = Never

Criterion	Rating
■ This team member is **honest** with other team members.	_____
■ This team member is **selfless** in putting the team's needs first.	_____
■ This team member's behavior promotes **unity** within the team.	_____
■ This team member **cooperates** in getting the job done.	_____
■ This team member works well with teammates of **diverse** backgrounds.	_____
■ This team member is **punctual.**	_____
■ This team member **perseveres** when the job is difficult.	_____
■ This team member **resolves conflict** in a positive manner.	_____
■ This team member's **attendance** meets expectations.	_____
■ This team member is **respectful** of others on the job.	_____
■ This team member applies **leadership** skills to help ensure a high level of team performance.	_____
■ This team member is an effective **communicator.**	_____
■ This team member seeks to build **trust** within the team.	_____
■ This team member is **dependable.**	_____
■ This team member takes **responsibility** for team performance.	_____
■ This team member takes the **initiative** to get the job done.	_____
■ This team member is **resourceful** in finding ways to get the job done.	_____
■ This team member is **tolerant** of differences in teammates.	_____
■ This team member is **supportive** of teammates in getting the job done.	_____
■ This team member is **aware** of the latest information that might affect the team's performance.	_____
■ This team member **participates** willingly in team activities.	_____
■ This team member is **courteous** to others on the job.	_____
■ This team member is willing to **follow** the team leader.	_____

FIGURE 3.4 Positive team behaviors.

ground rules, that might be included in a performance-appraisal form for team members.

It is important to note that these criteria cover only *positive team behaviors.* A comprehensive performance-appraisal form would also include criteria for assessing the individual's technical work skills. When considering the criteria in Figure 3.4, remember that the ground rules upon which they are based would have been developed with the input and concurrence of the team members.

The criteria shown in Figure 3.4 should be used to evaluate the performance of team members regardless of the type of team they serve on. The teamwork-oriented criteria may be included in a more comprehensive performance-appraisal form or be used separately. In the case of work teams, it is best to include the teamwork criteria in a more comprehensive form that also includes work-skills criteria. For ad hoc improvement teams and standing committees, the teamwork performance appraisal typically would be done separately. However, the results of a separate teamwork appraisal should still be used by the individual's supervisor when conducting his regular, periodic performance-appraisal. Tying together teamwork performance with all other aspects of on-the-job performance is critical.

PEER ACCOUNTABILITY

When it is done right, peer accountability can be the most effective form of accountability in teams. The more teammates hold each other accountable, the less supervision is required to get the job done. There are some excellent examples of teams in which peer accountability is the primary form of accountability—two of the best of which are found in the U.S. Marine Corps and the U.S. Navy SEALS. The most basic operational unit in the Marine Corps is the *fire team.* In the Navy SEALS, it is the *boat team.* Peer accountability is the norm in these teams, and it is very effective. Most members of a Marine fire team or a SEAL boat team would rather die than let their teammates down. Over the years, many have.

The development of the "we-are-in-this-together" attitude that makes peer accountability so effective in the Marine Corps and SEAL teams begins on the first day of training. Of course, the Marine Corps and the Navy have greater latitude in the ways they can develop team spirit than private companies do, but there is still much that can be learned from their example. Marine and SEAL trainees learn early that the entire team is disciplined every time even one member fails to perform properly. They also learn that success depends on teamwork and that a team can run, swim, hike, or climb only as fast as its slowest member—a tremendous incentive for the slower members to improve and for the faster members to help them do so.

Before becoming a Marine or a SEAL, trainees learn to depend on each other for even the most basic necessities (e.g., eating, sleeping, first aid). If they succeed, they succeed together. If they fail, they fail together. A supervisor in a technology company cannot tell his team members to drop to the floor and grind out pushups every time an individual member fails to perform, but he can still establish effective peer accountability.

TEAMWORK TIP

"A man watches himself best when others watch him too."

—George Savile, English politician

To establish effective peer accountability, team leaders in the workplace must:

- set a positive example of the behaviors expected of team members.
- reiterate what is expected of the team (especially the groundrules).
- talk openly and frequently about peer accountability.
- reinforce peer accountability.

Set a Positive Example

The most fundamental rule of leadership is to set a positive example for those you lead. The team leader who expects team members to be honest with each other must be scrupulously honest with them. The team leader who expects punctuality must be consistently punctual herself. Team members will not hold each other accountable for behavior their team leader fails to exhibit.

Reiterate the Expectations

Expectations for team members were established in the development of the ground rules. When attempting to establish peer accountability, the team leader must reiterate the expectations and remind team members that these are "*our*" ground rules. This can be done effectively in a team meeting: ground rules can be reviewed and discussed, and then the need for peer accountability can be introduced.

Talk About Peer Accountability

Team leaders' expectations should never be a secret. Team leaders should tell their team members that peer accountability is expected and talk with them about it. For team members, the opportunity to talk about peer accountability is an important prerequisite to accepting it. Some people will be uncomfortable at first because they do not like the confrontation that peer accountability may require. Others will balk because they do not want teammates holding their feet to the fire. These and other issues should be discussed openly. Team members need to understand that not only do they have a right to hold each other accountable, they have a responsibility to do so, because their performance depends on that of other team members. The basic message is this: "Because your performance affects my performance, pay, or advancement potential, I am going to hold you accountable. Correspondingly, I expect you to hold me accountable."

Reinforce Peer Accountability

When a team leader sees peer accountability at work, she should encourage it through positive reinforcement. This can take many forms. A public "way-to-go" or a simple pat on the back goes a long way toward encouraging the desired behavior. If disputes arise among team members, the team leader should settle the dispute in favor of peer accountability. If, for instance, a team member approaches her team leader and says, "I wish you would get these people off my back; I show up late a few times and they are on my case," the proper response is that "these people" are supposed to be on her case. They are trying to get her to pull her weight in the team. This situation would also provide a good opportunity for the team leader to explain how the work of the team is adversely affected by the employee's tardiness.

Summary

1. A good foundation for team accountability exists when the team knows what is expected of it, knows how successful fulfillment of expectations will be determined, has been given the training necessary to fulfill expectations, and is given the resources and support necessary to fulfill expectations.

2. Formal accountability involves written evaluations of the team's performance based on accomplishment of the team's charter and evaluations of team members based on adherence to the team's ground rules. Informal accountability involves ongoing monitoring by the team leader, monitoring by fellow team members (peer accountability), and self-monitoring by individual team members.

3. Accountability for team leaders is based on the team's performance in accomplishing the team's charter. Team leadership criteria should be built into the periodic performance appraisals of all individuals who lead teams as part of their official duties.

4. Accountability for team members begins with the team's ground rules—another reason why it is so important to develop a comprehensive set of team ground rules and to involve all team members in their development.

5. Peer accountability, when done right, can be the most effective form of accountability in teams. The more teammates hold each other accountable, the less supervision is required to get the job done. To establish effective peer accountability, team leaders must do four things: (1) set a positive example of the behaviors expected of team members, (2) reiterate what is expected of the team, (3) talk openly and frequently about peer accountability, and (4) reinforce peer accountability.

Key Terms and Concepts

Accountability
Accountability based on team performance
Accountability for team leaders
Accountability for team members
Formal accountability
Foundation for accountability
Informal accountability
Ongoing monitoring by the team leader
Ongoing peer monitoring
Peer accountability
Reinforce peer accountability
Reiterate the expectations
Set a positive example
Talk about peer accountability
Team leader's periodic performance appraisal

Review Questions

1. Why is it important that teams be accountable for their performance?
2. List the various elements of the foundation for team accountability.
3. Compare and contrast formal versus informal accountability.
4. Explain the concept of ongoing monitoring by the team leader.
5. Explain the concept of ongoing peer monitoring.
6. How should team leaders be held accountable?
7. How should team members be held accountable?
8. What is peer accountability, and why is it important?
9. How can a team leader establish effective peer accountability?

EFFECTIVE TEAMWORK SIMULATION CASES

The following cases deal with specific issues relating to the implementation of effective teamwork. Each case represents a meeting of Marcee McPhee and Pete

Fared, engineers and team leaders at Mac-Tech, Inc., a firm with 526 employees. McPhee and Fared are not just colleagues; they are friends, and their friendship goes all the way back to college. They both attended the same engineering school and graduated in the same class. On the job, their relationship has evolved into one of mentoring, in which McPhee is helping Fared learn to be a better team leader. Once a week they meet for lunch and discuss problems, progress, issues, and concerns. These cases chronicle their luncheon conversations and invite the reader to discuss the issues Fared and McPhee deal with.

CASE 3.1 Team Accountability

"Marcee, I think we are missing something in the development of teamwork in our company." "Oh really?" asked Marcee McPhee. "And what would that be?" McPhee felt sure she knew where the conversation was headed, but rather than interject her opinion, she sat back and listened. She wanted to get a better feel for Pete Fared's perspective before offering hers.

"Well, think about it, Marcee. We've given our teams a charter that explains the big picture and the goals to be accomplished. We've worked with them to develop ground rules so they know what is expected in terms of positive team behaviors. But there is still something missing." "You've been giving this a lot of thought haven't you?" asked McPhee. "I sure have. I woke up at three o'clock this morning and couldn't get back to sleep," said a fatigued Fared. "Pete, what do you think is the missing ingredient?"

Fared leaned back in his chair. After a few moments, he said, "The test." Confused, McPhee responded, "The test?" "Yes, the test. That's what's missing, Marcee. The test." "Help me out here, Pete. We've been talking about team charters and ground rules. Now, in mid-stream, you've changed horses on me." Then, before Fared could answer, she realized what he was talking about.

"Remember when we were in college, Marcee? In every course, our instructors gave us a syllabus that contained the learning goals, assignments, grading methods, testing dates, and other pertinent information about the course." "I remember," said McPhee.

"The syllabus explained what was expected of us," said Fared, warming to his lecture. "But the syllabus didn't hold us accountable. Expectations and measurement are two different things. That's where the tests came in. Without tests and other graded work, there would have been no accountability." McPhee smiled broadly and said, "Bingo."

Discussion Questions

1. What problems would occur if a college professor required assignments in her course, but gave no tests and graded no work?
2. Can a company have an effective team accountability system without measuring performance? Explain your opinion.

CASE 3.2 Some Progress Made on Team Accountability

"Now that my team has a charter and a good set of ground rules, our performance ratings are getting steadily better. In fact, our boss actually gave me a compliment this week," said Pete Fared. "Not bad," responded Marcee McPhee. "He doesn't hand out compliments freely. Congratulations Pete." Fared and McPhee were updating each other on progress and problems with their teams during their weekly luncheon meeting. "Well, since the boss appeared to be in a good mood, I took the opportunity to bring up the issue we discussed last week," related Fared. "And. . . " McPhee made a gesture that said, "Go on."

"Well, he didn't throw me out," offered Fared with a grin. "I guess that's something." "Come on Pete. Give. What did he say?" McPhee had been diligently swimming upstream for months trying to get company officials interested in establishing the necessary infrastructure to support teamwork. Finally, it looked like her colleague might have made some headway. If so, she was anxious to know about it. "He told me to rough out a draft of what I thought the company's performance-appraisal form should look like with regard to teamwork-positive behaviors," stated Fared. "Wait. He said that? He said '*teamwork-positive behaviors?*' " asked McPhee, incredulous. "That's what he said," acknowledged Fared. "Well I'm glad to hear it," said McPhee shaking her head.

Discussion Questions

1. Fared and McPhee work for a company that has not yet established teamwork as the way of doing business. Consequently, they keep encountering problems as they attempt to "lead from the bottom up." Put yourself in their shoes and discuss the types of problems you might encounter trying to apply leadership from the bottom.
2. What criteria would you include in the performance-appraisal form for team members if you were Fared and McPhee?

CASE 3.3 Establishing Peer Accountability Is Not Easy

"How'd it go with your team, Marcee?" Pete Fared began questioning his colleague even as he slid into a chair across the table from her. "Well Pete, we made some progress, but we still have a little work to do. We're not there yet. The thought of peer accountability makes some of my team members uncomfortable." Marcee McPhee stared past Fared into the distance as she said this, shaking her head slightly as if remembering. "How'd it go with your team?" asked McPhee. "About the same as yours, I guess. Some of my folks are a little nervous about it, too."

Fared and McPhee had called team meetings that morning to introduce the concept of peer accountability. Now that their employer, Mac-Tech, Inc., had bought into the idea of implementing teamwork companywide, all teams finally had charters and ground rules. Performance-appraisal forms for all employees were being revised at that very moment to include teamwork criteria.

Because of the outstanding work McPhee and Fared had done plowing new ground in the area of teamwork, their teams had been designated as "guinea pigs" for new teamwork concepts before they were implemented companywide. The latest concept was peer accountability.

"One of my team members said it seemed like I was trying to get the others to spy on him," said McPhee. "By the way Pete, it turns out that this same guy is late coming to work every time I have to attend our bi-weekly team leaders' meeting." "Let me see if I've heard you correctly, Marcee. Every time you have to be away for a meeting, this employee comes to work late, and now he is uncomfortable with peer accountability. I wonder why?" "I know," acknowledged McPhee. "It's always the guilty who protest the loudest."

"I have a couple team members who don't like the idea of peer accountability because they are afraid it might result in conflicts between team members," offered Fared. "I have one who feels that way, too," admitted McPhee. "How did you handle that issue, Pete?" Fared explained that he shared some stories from his days as a baseball player with his team members. In one story, he and his teammates collectively demanded that the coach bench the team's starting third baseman, usually one of the best players on the team. This player was making an uncharacteristically high number of errors and his batting average had plunged to an all-time low. Every player on the team knew the third baseman was staying out late at night and drinking heavily, but that wasn't the issue they went to the coach with. Their demand that he be benched was based on his poor performance on the field, not his after-hours activities.

According to Fared, the team's focus on the third baseman's performance rather than his nightlife was important. Although they did not approve of his after-hours activities, this player's drinking was his business. His performance on the baseball field, however, was their business. What he did on the field—or what he didn't do—affected them. That gave them the right to hold him accountable. "Did your story do any good?" asked McPhee. "Some, but I still have more work to do with my team." "Me too," said McPhee

Discussion Questions

1. Put yourself in the place of one of the team members in this case. How would you feel about peer accountability? Explain your reasoning.
2. Put yourself in the shoes of either Fared or McPhee. How would you respond to team members who object to peer accountability?

4

Develop Team-Leadership Skills

True leadership must be for the benefit of the followers, not the enrichment of the leader.

—Robert Townsend

OBJECTIVES

- Explain the concept of developing team leaders.
- Define "team leadership."
- Name common leadership myths, and explain why they are untrue.
- Describe the major categories of team-leadership styles.
- Explain how to build and maintain a following.
- Describe the responsibilities of team leaders in making ethical decisions.

Leadership is the single most important ingredient in achieving consistent peak performance in teams. With good leadership, ordinary people can achieve extraordinary results. On the other hand, even the most talented people will produce mediocre results if they are poorly led. This chapter explains how to develop leadership skills in those responsible for leading teams.

DEVELOPING TEAM LEADERS

Perhaps the oldest debate about leadership revolves around the question, "Are leaders born or made?" Can leadership skills be learned, or must they be inherited? This debate hasn't yet been settled, and it probably never will be. There are proponents of both sides of the debate, and this polarity is not likely to change because, as often is the case in such controversies, both sides are partially right.

Leaders are like athletes: some athletes are born with natural ability, and others develop their ability through determination and hard work. Inborn ability, or the lack thereof, represents only the starting point. Success from that point forward depends on the individual's willingness and determination to develop and improve. Some athletes born with tremendous natural ability never live up to their potential. Other athletes with limited natural ability perform beyond their apparent potential. Similarly, some managers have more natural leadership ability than others. Regardless of their individual starting points, however, all managers can become good leaders through education, training, practice, determination, and effort.

Technology companies can help their key personnel develop leadership skills by providing the necessary training and mentoring. Leadership training can be done in-house or externally—by bringing in a trainer or by sending key personnel to programs provided by colleges, universities, or private-training companies. The necessary training can also be provided through online courses.

Regardless of the approach, technology companies should ensure that certain essential information is covered in the training. The remainder of this chapter contains the information that should be covered in a leadership training program.

TEAMWORK TIP

"You cannot manage people into battle. You manage things; you lead people."

—Grace Murray Hopper, Admiral, United States Navy

WHAT IS TEAM LEADERSHIP?

What Team Leaders Must Be Able to Do

In their book *Leaders: The Strategies for Taking Charge,* Warren Bennis and Burt Nanus describe three lessons for leadership that summarize what leaders must be able to do.[1]

- *Overcome resistance to change.* Some people attempt to do this using power and control. Leaders overcome resistance by achieving a total, willing, and voluntary commitment to shared values and goals.
- *Broker the needs of constituency groups inside and outside of the team.* When the needs of the team and another unit appear to conflict, leaders must be able to find ways of bringing the needs of both together without shortchanging either one.
- *Establish an ethical framework within which all team members and the team as a whole operate.* This is best accomplished by doing the following:
    - Setting an example of ethical behavior
    - Choosing ethical people as team members
    - Communicating a sense of purpose for the team
    - Reinforcing appropriate behaviors within the team and outside of it
    - Articulating ethical positions, internally and externally

What Is a Good Team Leader?

Good team leaders come in all shapes, sizes, genders, ages, races, political persuasions, and national origins. They do not look alike, talk alike, or even work alike. However, good leaders do share several common characteristics (see Figure 4.1). These characteristics inspire people to make a total, willing, and voluntary commitment.

Good leaders are committed both to the job to be done and to the people who must do it, and they are able to strike the appropriate balance between the two. Good leaders project a positive example at all times. They are good role models. Managers who project a "Do as I say, not as I do" attitude are not effective team leaders. To inspire employees, team leaders must be willing to do what they expect of workers—and to do it better, do it right,

Checklist of Characteristics of Good Team Leaders

_✓_ Balanced commitment between the work and team members.

_✓_ Positive role model for team members.

_✓_ Good communication skills.

_✓_ Positive influence on team members.

_✓_ Persuasiveness with team members.

FIGURE 4.1 Good team leaders have these personal characteristics.

and do it consistently. If, for example, dependability is important, team leaders must set a consistent example of dependability. If punctuality is important, team leaders must set a consistent example of punctuality. To be a good leader, one must set a consistent example of all characteristics that are important on the job.

Good leaders are good communicators. They are willing, patient, skilled listeners. They are also able to communicate their ideas clearly, succinctly, and in a nonthreatening manner. They use their communication skills to establish and nurture rapport with team members. Good leaders have influence with employees and use it in a positive manner. *Influence* is the art of using power to move people toward a certain end or point of view. The power of team leaders derives from the authority that goes with their jobs and the credibility they establish. Power is useless unless it is converted to influence. Power that is properly, appropriately, and effectively applied becomes positive influence.

Finally, good leaders are persuasive. Team leaders who expect people simply to do what they are told to do have limited success. Those who are able to use their communication skills and influence to persuade people to their point of view and to convince them to make a total, willing, and voluntary commitment to that point of view can have unlimited success.

LEADERSHIP MYTHS

Over the years, a number of myths have grown up about leadership. Leaders in a team setting should be aware of these myths and be able to dispel them. Bennis and Nanus describe the most common myths about leadership as follows.[2]

Myth: Leadership Is a Rare Skill

Although it is true that there are few great leaders of world renown, there are many good, effective leaders. Renowned leaders such as Winston Churchill were good leaders given the opportunity to participate in monumental events. General Norman Schwarzkopf is another example. He had always been an effective military leader; that's how he became a general. But it took a monumental event—the Gulf War—coupled with that leadership ability to make him a world-renowned leader. His leadership skills didn't appear suddenly; he had them all along. Circumstances allowed them to be displayed on the world stage.

Most effective leaders spend their careers in virtual anonymity, but they exist in surprisingly large numbers. There may be little or no correlation between their ability to lead and the level of their positions in an organization. The best leader in a company may be the lowest-paid wage earner, and the worst may be the CEO. In addition, a person may be a leader in one setting and not in another. For example, a person who shows no leadership ability at work may be an effective leader in his church. One of the keys to success for companies that must compete in the global marketplace is to create an environment that brings out the leadership skills of all employees at all levels and focuses these skills on continually improving performance.

Myth: Leaders Are Born, Not Made

Leadership attitudes and behaviors can be learned, even by those who do not appear to have innate leadership potential.

Myth: Leaders Are Charismatic

Some leaders have charisma and some don't. Some of history's most renowned leaders have had little or no charisma. Correspondingly, some of history's greatest misleaders have been highly charismatic. Generals Dwight Eisenhower and Omar Bradley are examples of great but uncharismatic leaders. Adolf Hitler and Benito Mussolini are examples of great misleaders who relied almost exclusively on charisma to build a following.

Myth: Leadership Exists Only at the Top

In reality, the opposite of this myth is often true. Top managers may be the least capable leaders in a company. Leadership relates to producing results and generating continual improvement, not one's relative position within the organization. Organizational competitiveness relies on building teams at all levels and teaching employees in these teams to be leaders.

Myth: Leaders Control, Direct, Prod, and Manipulate

If practice is an indicator, this myth is the most widely believed. The "I'm the boss, so do what I say" syndrome is rampant in business and industry. It seems to be the automatic default approach for managers who don't know better. Leadership in a team setting is about involving and empowering, not prodding and manipulating.

Myth: Leaders Don't Need to Be Learners

Lifelong learning is a must for team leaders. Leaders don't learn simply for the sake of learning (although doing so is a worthwhile undertaking). Rather, leaders continually learn in an organizational context. This means they approach learning from the perspective of what matters most to their organizations. A manager who is responsible for the metal-fabrication team in a manufacturing firm might undertake to learn more about the classics of European literature. Although this would certainly make her a better-educated person, studying European literature does not represent learning in an organizational context for this manager. More appropriate would be techniques to improve speed and feed rates, statistical process control (SPC), team-building strategies, computer numerical control programming, information about new composite materials, total productive maintenance, or anything else that would help improve the team's performance.

TEAMWORK TIP

"Some are born great, some achieve greatness, and some have greatness thrust upon them."

—William Shakespeare

TEAM LEADERSHIP STYLES

Leadership styles have to do with how leaders interact with those they lead. Leadership styles go by many different names, but most fall into one of the following categories: autocratic, democratic, participative, goal-oriented, and situational.

Autocratic Leadership

Autocratic leadership is also called *directive* or *dictatorial leadership*. People who take this approach make decisions without consulting the employees who will have to implement them or who will be affected by them. They tell others what to do and expect them to comply. Critics of this style say that although it can work in the short run or in isolated instances, it is not effective in the long run. Autocratic leadership is not appropriate in a team setting.

Democratic Leadership

Democratic leadership is also called *consultative* or *consensus leadership*. People who take this approach involve in decision making the employees who will have to implement the decisions. The leader makes the final decision, but he does so only after receiving the input and recommendations of team members. Critics of this approach say the most popular decision is not always the best decision and that democratic leadership, by its nature, can result in popular decisions as opposed to right decisions. This style can also lead to compromises that ultimately fail to produce the desired result. Democratic leadership is not appropriate in a team setting.

Participative Leadership

Participative leadership is also known as *nondirective leadership*. People who take this approach exert reduced control over the decision-making process. They provide information about the problem and ask team members to develop strategies and solutions. The underlying assumption of this style is that workers more readily accept responsibility for solutions, goals, and strategies if they help develop them. Critics of this approach say it is time consuming and works only if all people involved are committed to the best interests of the team.

Goal-Oriented Leadership

Goal-oriented leadership is also called *results-based leadership*. People who take this approach ask team members to focus solely on the goals at hand. Only those strategies that make a definite and measurable contribution to accomplishing team goals are discussed. The influence of personalities and other factors unrelated to the specific goals of the team are minimized. Critics of this approach say it can break down when team members focus so intently on specific goals that they overlook opportunities or potential problems that fall outside their narrow focus. These drawbacks can make results-based leadership ineffective in team settings.

Situational Leadership

Situational leadership is also called *contingency leadership*. People who take this approach select the leadership style that seems appropriate based on the circumstances that exist at a given time. In evaluating these circumstances, leaders consider the following factors:

- The relationship of the leader and team members
- How precisely actions must comply with specific guidelines
- The amount of authority the leader actually has with team members

Depending on these factors, the manager decides whether to take the autocratic, democratic, participative, or goal-oriented approach. Under different circumstances, the same leader would apply a different leadership style. Detractors reject situational leadership as an attempt to apply an approach based on short-term concerns instead of on the solution of long-term problems.

Best Leadership Style in a Team Setting

The most appropriate leadership style in a team setting might be called *participative leadership taken to a higher level*. Whereas participative leadership in the traditional sense involves soliciting employee input, in a team setting it involves soliciting input from empowered employees, listening to that input, and acting on it. The key difference between traditional participative leadership and participative leadership from a team perspective is that, with the latter, the employees providing input are empowered to take initiative in solving problems and making improvements.

Collecting employee input is not new. However, collecting input, logging it in, tracking it, acting on it in an appropriate manner, working with employees to improve weak suggestions rather than simply rejecting them, and rewarding employees for improvements that result from their input extend beyond the traditional approach to participative leadership.

BUILDING AND MAINTAINING A FOLLOWING

People can be leaders only if those they hope to lead follow them willingly and steadfastly. Followership must be built, and once built, maintained. This section is devoted to a discussion of how team leaders build and maintain followership among the people they hope to lead.

Popularity and the Leader

Many people confuse popularity with leadership and, in turn, followership. Leadership and popularity are not the same thing, however. Long-term followership grows out of respect, not popularity. Good leaders *may* be popular, but they *must* be respected. Not all good leaders are popular, but all are respected.

Team leaders occasionally have to make unpopular decisions. This is a fact of life, and it explains why leadership positions are sometimes described as lonely. Making an unpopular decision does not necessarily cause a leader to lose followership, provided she is seen as having solicited a broad base of input and having given serious, objective, and impartial consideration to that input. Correspondingly, leaders who make inappropriate decisions that are popular in the short run may actually lose followership in the long run. If the long-term consequences of a decision turn out to be detrimental to the team, team members will hold the leader responsible, especially if the decision was made without first collecting and considering employee input.

Leadership Characteristics That Build and Maintain Followership

Team leaders build and maintain followership by earning the respect of those they lead. Here are some characteristics of leaders that build respect (see also Figure 4.2).

Checklist of Characteristics That Build and Maintain Followership

- ✓ Sense of purpose
- ✓ Self-discipline
- ✓ Honesty
- ✓ Credibility
- ✓ Common sense
- ✓ Stamina
- ✓ Commitment
- ✓ Steadfastness

FIGURE 4.2 Team leaders build and maintain followership by displaying these characteristics.

- *Sense of purpose.* Successful leaders have a strong sense of purpose. They know who they are, where they fit into the overall organization, and the contributions their areas make to the organization's success.
- *Self-discipline.* Successful leaders develop discipline. Self-discipline allows leaders to avoid negative self-indulgence, inappropriate displays of emotion (such as anger), and counterproductive responses to the everyday pressures of the job. Through self-discipline, leaders set an example of handling problems and pressures with equilibrium and a positive attitude.
- *Honesty.* Successful leaders are trusted by their followers. They are open, honest, and forthright with team members and with themselves. They can always be depended on to make difficult decisions in unpleasant situations.
- *Credibility.* Successful leaders have credibility, established by being knowledgeable, consistent, fair, and impartial in all human interactions; setting a positive example; and adhering to the same standards of performance and behavior expected of others.
- *Common sense.* Successful leaders know what is important in a given situation and what is not. They know that tact is important when dealing with people. They know when to be flexible and when to be firm.
- *Stamina.* Successful leaders must have stamina. Frequently, they need to be the first to arrive and the last to leave. Their hours are

likely to be longer and the pressures they face more intense than those of others. Energy, endurance, and good health are important to those who lead.

- *Commitment.* Successful leaders are committed to the goals of the team, the people they work with, and their own ongoing personal and professional development. They are willing to do everything within the limits of the law, professional ethics, and company policy to help their team succeed.
- *Steadfastness.* Successful leaders are steadfast and resolute. People will not follow a person they perceive to be wish-washy and noncommittal, or whose resolve they question. Successful leaders must have the steadfastness to stay the course even when it is difficult.

Pitfalls That Can Undermine Followership

The previous section put forth positive characteristics that help team leaders build and maintain the respect and, in turn, the followership of those they hope to lead. Leaders should also be aware of several common pitfalls that can undermine followership and the respect they work so hard to earn.

- *Trying to be a buddy.* Positive relations and good rapport are important, but leaders are not the buddies of those they lead. The nature of the relationship does not allow it.
- *Having an intimate relationship with a team member.* This practice is both unwise and unethical. A positive leader–employee relationship cannot exist in this circumstance. Few people can succeed at being the lover and the leader, and few things can damage the morale of a team as quickly and completely.
- *Trying to keep things the same when leading former peers.* The leader–employee relationship, no matter how positive, is different from the peer–peer relationship. This can be a difficult adjustment to make, but it must be made if the peer-turned-team leader is to succeed.

ETHICS IN TEAM LEADERSHIP

Although there are many definitions of the term "ethics," no one definition has been universally accepted. The often conflicting and contradictory interests of workers, customers, competitors, and the general public result in a propensity for ethical dilemmas in the workplace. Ethical dilemmas in the workplace are often more complex than other ethical situations. They involve social expectations, competition, and responsibility as well as the potential consequences of an employee's behavior to the organization's many constituencies.

In the discussion of ethics, the terms "conscience," "morality," and "legality" are frequently heard. Although these terms are closely associated with ethics, they do not by themselves define it. For the purposes of this book, ethics is defined as follows:

> ***Ethics** is the application of morality. **Ethical behavior** means doing the "right thing" within a moral framework.*

"Morality" refers to the values that are widely subscribed to and fostered by society in general and individuals within society. The field of ethics attempts to apply reason in determining rules of human conduct that translate morality into everyday behavior. Ethical behavior is that which falls within the limits prescribed by morality. Doing what is ethical is often called doing the "right thing."

How, then, does a team leader know if someone's behavior is ethical? Ethical questions are rarely black and white. They typically fall into a gray area between the two extremes of clearly right and clearly wrong; this gray area is often further clouded by personal experience, self-interest, point of view, and external pressure.

GUIDELINES FOR DETERMINING ETHICAL BEHAVIOR

Before presenting guidelines team leaders can use in sorting out matters that fall into the gray area between clearly right and clearly wrong, it is necessary to distinguish between "legal" and "ethical." They are not always synonymous.

It is not uncommon for people caught in the practice of questionable behavior to use the "I didn't do anything illegal" defense. Behavior can be well within the scope of the law and still be unethical. The following guidelines for determining ethical behavior assume that the behavior is legal:

- Apply the *morning-after test.* If you make this choice, how will you feel about it and yourself tomorrow morning?
- Apply the *front-page test.* If it was printed as a story on the front page of your local newspaper, how would you feel about your decision?
- Apply the *mirror test.* If you make this decision, how will you feel about yourself when you look in the mirror?
- Apply the *role-reversal test.* Mentally trade places with the people affected by your decision. How does it look through their eyes?
- Apply the *common-sense test.* Listen to what your instincts tell you. If it feels wrong, it probably is.

THE TEAM LEADER'S ROLE IN ETHICS

Using the guidelines set forth in the previous section, team leaders should be able to make responsible decisions concerning ethical choices. Unfortunately, deciding what is ethical is often easier than *doing* what is ethical. In this regard, trying to practice ethics is like trying to diet. It is not just a matter of knowing you should cut down on your caloric intake, it is also a matter of following through and actually doing it.

It is this fact that defines the role of team leaders with regard to ethics. Team leaders have a three-part role. First, they are responsible for setting an example of ethical behavior. Second, they are responsible for helping employees make the right decisions when facing ethical questions. Third, team leaders are responsible for helping employees follow through on the ethical option once the appropriate choice has been identified. In carrying out these responsibilities, team leaders can adopt one of the following approaches:

- Best-ratio approach
- Black and white approach
- Full potential approach

Best-Ratio Approach

The best-ratio approach is the pragmatic option. Its philosophy is that people are basically good and under the right circumstances, they will behave ethically. However, under certain conditions they can be driven to unethical behavior. Therefore, the team leader should create conditions that promote ethical behavior and maintain the best possible ratio of good choices to bad. When hard decisions must be made, team leaders should make the choice that will do the most good for the most people. This is sometimes referred to as "situational ethics."

Black-and-White Approach

In the black-and-white approach, right is right and wrong is wrong. Circumstances and conditions are irrelevant. The team leader's job is to make ethical decisions, carry them out, and help employees choose the ethical route regardless of circumstances. When difficult decisions must be made, leaders should make fair and impartial choices and let the chips fall where they may.

Full-Potential Approach

Under the full-potential approach, team leaders make decisions based on how they will affect the ability of the people involved to achieve their full potential. The underlying philosophy is that people are responsible for realizing their full potential within the confines of morality. Choices that achieve this goal without infringing on the rights of others are considered ethical.

Summary

1. The age-old debate concerning whether leaders are born or made continues. The point of view set forth in this book is that *leaders are like athletes.* Some are born with natural ability, and some develop their ability through hard work and determination. Innate ability represents only the starting point. Regardless of their innate ability, people can become good leaders through training, practice, determination, and effort.

2. Team leadership is the ability to inspire team members to make a total and willing commitment to accomplishing or exceeding the team's goals. A team leader must be able to overcome resistance to change, broker the needs of constituency groups inside and outside of their team, and establish an ethical framework within which the team must operate.

3. Leadership myths that have been perpetuated over the years include the following: leadership is a rare skill; leaders are born, not made; leaders are charismatic; leadership exists only at the top; and leaders control, direct, and prod.

4. Common leadership styles include the following: autocratic, democratic, participative, goal-oriented, and situational. The best style for teams is participative leadership taken to a higher level.

5. Personal characteristics that help team leaders build a following include: a sense of purpose, self-discipline, honesty, credibility, common sense, stamina, commitment, and steadfastness.

6. Pitfalls that can undermine leadership include: trying to be a buddy, having an intimate relationship with a team member, and trying to keep things the same when leading former peers.

7. Ethics is the application of morality. Ethical behavior means doing the right thing within a moral framework. Tests for determining the ethical course of action include: the morning-after test, the front-page test, the mirror test, the role-reversal test, and the common-sense test.

8. When making ethical decisions, team leaders may adopt one of the following approaches: best-ratio, black and white, and full potential.

Key Terms and Concepts

Autocratic leadership
Best-ratio approach
Black-and-white approach
Commitment
Common sense
Common-sense test
Credibility
Democratic leadership
Developing team leaders
Ethics

Followership
Front-page test
Full-potential approach
Goal-oriented leadership
Good team leader
Honesty
Leaders are born not made(?)
Leaders are charismatic(?)
Leaders control, direct, prod, and manipulate(?)
Leaders don't need to be learners(?)
Leadership
Leadership exists only at the top(?)
Leadership is a rare skill(?)
Mirror test
Morning-after test
Participative leadership
Role-reversal test
Self-discipline
Sense of purpose
Situational leadership
Stamina
Steadfastness

Review Questions

1. Defend or refute the following statement: Leaders are born, not made.
2. Define "leadership" as it applies to teamwork.
3. What three things must leaders be able to do?
4. What are the characteristics of a good leader?
5. List and explain four myths about leadership.
6. List and explain five different leadership styles.
7. What is the best leadership style in the context of a team? Why?
8. Defend or refute the following statement: Leaders must be popular with those they hope to lead.
9. List and explain the leadership characteristics that build and maintain followership.
10. What are three pitfalls that can undermine team leadership?
11. How do ethics apply to the teamwork situation?
12. List and explain five tests that can be used when facing an ethical dilemma.
13. How would you describe the team leader's role in ethics?

EFFECTIVE TEAMWORK SIMULATION CASES

The following cases deal with specific issues relating to the implementation of effective teamwork. Each case represents a meeting of Marcee McPhee and Pete Fared, engineers and team leaders at Mac-Tech, Inc., a technology firm with 526 employees. McPhee is the leader of Team A, and Fared leads Team B. McPhee and Fared are not just colleagues; they are friends, and their friendship goes all the way back to college. They both attended the same engineering school and graduated in the same class. On the job, their relationship has evolved into one of mentoring, in which McPhee is helping Fared learn to be a better team leader. Once a week they meet for lunch and discuss problems, progress, issues, and concerns. These cases chronicle their luncheon conversations and invite the reader to discuss the issues Fared and McPhee deal with.

CASE 4.1 Are Leaders Born or Made?

"What do you think about this new leadership-training program the boss is requiring?" asked Marcee McPhee. "I'll tell you what I think," said Pete Fared emphatically. "I think you and I have created a monster." For months McPhee had tried unsuccessfully to persuade their boss of the potential benefits of teamwork. Finally, out of frustration she had enlisted the help of Fared. When Fared approached their boss, the timing had been right. Suddenly the manager who had ignored McPhee's attempts to persuade him of the need for teamwork in their company had become an advocate of the concept. Now he was showing so much enthusiasm that McPhee and Fared found themselves alternating between incredulity and joy.

"You should be happy, Pete. Now that we have an executive-level manager on our side, things are beginning to happen. I swam upstream against a strong current for six months trying to generate some interest in teamwork. Now I can barely keep up with all that is happening." Fared nodded to acknowledge his colleague's justifiable relief and replied, "I know Marcee, and I'm pleased too. It's not the hard-won support we are finally getting that I am complaining about. It's the leadership training."

McPhee was clearly surprised at his reservations. "What's bothering you about the training, Pete? I think the trainer is doing a good job. Don't you?" Fared explained that he, too, was impressed with the trainer. His problem was with the concept of leadership training in general. "I'm just not convinced you can train people to be leaders, Marcee. I think you have to be born with the characteristics the trainer talked about in our first session. Think about it, Marcee. Do you really believe people can learn common sense, honesty, fairness, self-discipline, and all of those other characteristics the trainer said are the foundation of good leadership?"

Tapping the table for emphasis, McPhee countered, "Yes, I believe most people can learn those things, Pete. Just look at honesty and fairness, for example. People who have those characteristics learned them from their parents or a teacher or a coach. Kids aren't naturally fair or honest, and you know it, Pete." Fared smiled sheepishly. Thinking back on his childhood, he had to admit that his colleague had a good point, but he wasn't ready to concede the argument just yet. "All right then, Marcee, what about Abraham Lincoln? I doubt he ever attended a seminar on leadership, but look at his record. Who but Lincoln had the leadership ability to hold our country together during the Civil War? Lincoln was a born leader," claimed Fared triumphantly.

"Good example," acknowledged McPhee. "Let's look at Lincoln. By studying his example, we can learn the importance of using persuasion rather than coercion. We can learn the importance of having a goal and persevering until we achieve it, and we can learn that you sometimes have to take a lot of criticism and just keep going."

Discussion Questions

1. Fared thinks leaders are born. McPhee thinks leaders can be made. What is your opinion, and why?
2. Think of a good leader whose example supports your opinion in this debate. Give examples that explain your choice.

CASE 4.2 Team Leader or Team Dictator?

"I'm not sure I buy what that leadership trainer is trying to sell," said Pete Fared to his colleague, Marcee McPhee. That morning, Fared and McPhee had attended a two-hour session as part of a leadership-training program required of all team leaders in their company. "Is there anything in particular the trainer said that is bothering you, Pete?" "You bet there is," acknowledged Fared. He went on to explain that the concept of participative leadership didn't sit right with him. "Marcee, when I played high-school baseball and when I was in the military, leaders led and the rest followed. There wasn't any of this empowerment nonsense."

McPhee rolled her eyes and made a face that said, "Here we go again." "Pete, we've had this conversation before," she said. "In fact, I think I remember you saying there were times when you were in the Navy that a certain officer would have made better decisions if he had just bothered to ask your opinion." "There were times like that, but there were also times when that same officer didn't have the luxury of asking for opinions. He had only enough time to act, and when he did, he expected us to follow his lead. When it's crunch time, and I need my team members to do something, I don't want to have to waste time asking for opinions."

"You'd make a great team dictator, Pete, but you're missing the whole point about participative leadership," said McPhee. "The time to ask for employee input is before you are in a crisis situation. No leader calls for a discussion session when it's time to act and act fast. Can't you see that?" Fared shrugged and said, "Well . . . "

Discussion Questions

1. Join this debate about leadership styles. Do you agree with Fared or McPhee? Why?
2. Would you rather be led by a person who wants your participation or by someone who simply dictates? Why?

CASE 4.3 Ethical Dilemmas Are Common for Team Leaders

"I really enjoyed our team leadership session this morning," announced Pete Fared. "Well, that's a pleasant surprise, Pete. Up to now you've complained about every session we've attended," Marcee McPhee reminded her colleague. "I know, I know," acknowledged Fared with a dismissive wave of his hand. "But this morning's session had a lot of meat in it. I run into ethical dilemmas all the time. The trainer finally gave me something I can use: some models for making ethical decisions."

"I agree," said McPhee. "Ethical dilemmas seem to pop up in my team all the time." "Really? Do you mind giving me an example of one?" "Not at all. In fact, I was hoping we could discuss a dilemma I'm dealing with right now. I'll just leave out the names. Is that alright with you?" "No problem," said Fared. "That's a good idea."

McPhee explained that she had a single mother in her team who was one of her most productive and most popular team members. But there was a problem: this employee was showing up for work late or leaving early at least three days out of five. She had a sick daughter who needed her attention, which was the reason behind her attendance problems. Her team members were supportive and were covering for her. "What's the problem then?" asked Fared. "If her team members are covering for her, that's what teamwork is all about."

"I know, Pete. The ethical dilemma has to do with this employee's pay. Her teammates are punching in and punching out for her on the time clock so she doesn't lose any pay. She is being paid for time she hasn't worked, and although I'm sympathetic about her daughter and know she needs the money for hospital bills, the situation is making me very uncomfortable." "That's a tough one, Marcee. What are you going to do?" "I don't know. What would you do, Pete?"

Discussion Questions

1. Put yourself in Marcee McPhee's position. What would you do? Why?
2. Which model would you use in making this ethical decision? Why?

Endnotes

1. W. Bennis and B. Nanus, *Leaders: The Strategies for Taking Charge* (New York: Harper & Row, 1985), pp. 184–186.
2. Bennis and Nanus, *Leaders*, pp. 222–226.

5

Develop Communication Skills

I don't care how much a man talks, if he says it in only a few words.

—Josh Billings

OBJECTIVES

- Explain the importance of developing communication skills.
- Define "communication" and "effective communication."
- Name and explain the four components of the communication process.
- Explain the common inhibitors of communication.
- Explain the concept of listening as a communication tool.
- Explain the concept of nonverbal communication.
- Explain the concept of verbal communication.
- Explain how to communicate corrective feedback.
- Explain how to improve communication in teams.

Of the skills needed by team leaders and team members, communication is one of the most important. Communication is fundamental to leadership, motivation, problem solving, performance appraisal, ethics, discipline, training, mentoring, and all other areas of concern to team leaders and team members. This chapter explains how to develop communication skills in team leaders and team members.

DEVELOPING COMMUNICATION SKILLS

Communication may be the most imperfect of all human processes. This is because the quality of communication is affected by so many different factors (*e.g.*, speaking ability; hearing ability; language barriers; differing perceptions or meanings based on age, gender, race, nationality, and culture; attitudes; nonverbal cues; level of trust between sender and receiver). Because of these and other factors, effective communication is difficult at best. However, it is essential to the continual improvement of a team's performance. Communication is the oil that lubricates the machinery of human interaction.

Communication skills can be learned, which is fortunate indeed for technology companies and technical professionals. With sufficient training and practice, most people—regardless of their innate capabilities—can learn to communicate well.

COMMUNICATION DEFINED

Inexperienced team leaders often confuse *telling* with *communicating*. When a problem develops, they are likely to protest, "But I told him what to do." Inexperienced team leaders also often confuse *hearing* with *listening*. They are likely to say, "This isn't what I told you to do. I know you heard me. You were standing right next to me!"

In both cases, the leader has confused telling and hearing with communicating. This point is illustrated by a quotation attributed to less-than-forthright politicians: "I know you believe you understand what you think I said, but I am not sure you realize that what you heard is not what I meant."

This amusing quote does make a point. What you say is not necessarily what the other person hears, and what the other person hears is not necessarily what you intended to say. The key word is "understand." Communication may involve telling, but it is not *just* telling. It may involve hearing, but it is not *just* hearing. For the purposes of this book, communication is defined as follows:

> ***Communication*** *is the transfer of information that is received and fully understood from one source to another.*

A message can be sent by one person and received by another, but until the message is fully understood, there is no communication. This applies to spoken, written, and nonverbal messages.

Communication versus Effective Communication

When information conveyed is received and understood, communication has occurred. Understanding by itself does not necessarily make effective communication, however. *Effective communication* occurs when the information that is received and understood is acted on in the desired manner.

For example, a team leader might ask her team members to arrive at work fifteen minutes early for the next week to ensure that an important order goes out on schedule. All of the team members verify that they understand both the message and the reasons behind it. Without informing the leader, however, two team members decide they are not going to comply. This is an example of ineffective communication. The two nonconforming employees understood the message but decided to ignore it. The leader in this case failed to achieve acceptance of the message. Consequently, the communication was ineffective.

Effective communication is a higher level of communication. Because it implies understanding *and* acceptance, it requires persuasion, motivation, monitoring, and leadership.

TEAMWORK TIP

"Think like a wise man, but communicate in the language of the people."

—William Butler Yeats

Communication Levels

Communication takes place at several levels in an organization, as listed in Figure 5.1.

Although team leaders are actively engaged primarily in the first two levels, there may be occasions in which they are involved in the other levels, at least indirectly. Consequently, team leaders should be familiar with all four levels of communication.

1. *One-on-one-level communication* is just what the name implies—one person communicating with another. This might involve face-to-face conversation, a telephone call, an email message, or even a simple gesture or facial expression.
2. *Team* or *unit-level communication* is communication within a peer group. The primary difference between one-on-one and team communication is that, with the latter, all team members are

involved in the process at once, as in a team meeting to solve a problem or set goals.

3. *Organization-level communication* is communication among groups. A meeting involving the sales department, design department, and a production team would represent an opportunity for organization-level communication.
4. *Community-level communication* occurs among groups inside an organization and groups outside the organization. Perhaps the most common example is an interaction between a company's representative and newspaper, radio, or television reporters.

Levels of Communication

✓ One-on-one level
✓ Team or unit level
✓ Organization level
✓ Community level

FIGURE 5.1 Communication can occur at different levels.

COMMUNICATION AS A PROCESS

Communication is a process that involves several components: the *sender*, the *receiver*, the *medium*, and the *message* itself. The sender is the originator or source of the message. The receiver is the person or group for whom the message is intended. The message is the information that is conveyed, understood, accepted, and acted on. The medium is the vehicle used to convey the message.

There are three basic categories of mediums: *verbal*, *nonverbal*, and *written*. Verbal communication includes face-to-face conversation, telephone conversation, speeches, public address announcements, press conferences, and other means of conveying the spoken word. Nonverbal communication includes gestures, facial expressions, voice tone, and poses. Written communication includes letters, email, memoranda, billboards, bulletin boards, manuals, books, and any other means of conveying the written word.

Technological developments are significantly affecting our ability to convey information. These developments include word processing, satellite communication, computer modems, cordless telephones, cellular telephones, answering machines, facsimile machines, pocket-size dictation machines, email, and the Internet.

COMMON INHIBITORS OF COMMUNICATION

As advanced as communication-enhancing devices have become, there are still several inhibitors of effective communication (see Figure 5.2). Team leaders should be familiar with these inhibitors and be able to avoid or overcome them.

Differences in meaning. Differences in meaning are inevitable in communication, because we all have different backgrounds and levels of education. We might also come from different cultures, races, or countries. As a result, words, gestures, and facial expressions can have altogether different meanings to different people. To overcome this inhibitor, team leaders must invest the time to get to know their team members.

Insufficient trust. Insufficient trust can inhibit effective communication. If receivers do not trust senders, they may be overly sensitive or guarded. They might concentrate so hard on reading between the lines for a "hidden agenda" that they miss the real message. This is why trust building between team leaders and members is so important. It is well worth the time and effort.

Information overload. Because of advances in communication technology and the rapid and continual proliferation of information, we often find ourselves with more information than we can process effectively. This is known as "information overload," and it can actually cause a breakdown in communication. Team leaders can guard against information overload by screening, organizing, summarizing, and simplifying the information they convey to team members.

Checklist of Common Inhibitors of Effective Communication

- ✓ Differences in meaning
- ✓ Insufficient trust
- ✓ Information overload
- ✓ Interference
- ✓ Condescending tones
- ✓ Listening problems
- ✓ Premature judgments
- ✓ Inaccurate assumptions
- ✓ Technological glitches

FIGURE 5.2 These factors can decrease the effectiveness of communication.

Interference. Interference is any external distraction that prevents effective communication. This might be something as simple as background noise or as complex as atmospheric interference with satellite communications. Regardless of its nature, interference either distorts or completely blocks the message. Team leaders must be attentive to the environment when they plan to communicate with employees.

Condescending tones. Problems created by condescention result from the tone rather than the content of the message. People do not like to be talked down to, and team leaders should never do so.

Listening problems. Listening problems are one of the most serious inhibitors of effective communication. They can result both from the sender not listening to the receiver and vise versa. An entire section is devoted to listening later in this chapter.

Premature judgments. Premature judgments by either the sender or the receiver can inhibit effective communication. This is a type of listening problem, because as soon as we make a quick judgment, we are prone to stop listening. One cannot make premature judgments and maintain an open mind. Therefore, it is important for team leaders to listen nonjudgmentally when communicating with employees.

Inaccurate assumptions. Our perceptions are influenced by our assumptions; consequently, inaccurate assumptions can lead to inaccurate perceptions. Here's an example. Jeff Smith, a technician, has been taking off an inordinate amount of time from work lately. His team leader, Mary Jones, assumes Jeff is goldbricking. As a result, whenever Jeff makes a suggestion in a team meeting, Mary assumes he is just lazy and is suggesting the easy way out.

It turns out that Mary's assumption is inaccurate. Jeff is actually a highly-motivated, highly-skilled worker. His excessive time off is the result of a problem he is having at home, a problem he is too embarrassed to discuss. Because of an inaccurate assumption, Mary is missing out on the suggestions of a highly-motivated, highly-skilled employee. In addition, her misperception points to a need for building trust. Perhaps if Jeff trusted Mary more, he would be less embarrassed to discuss his personal problem with her.

Technological glitches. Software bugs, computer viruses, dead batteries, power outages, and software conversion problems are a few technological glitches that can interfere with communication. The more dependent we become on technology in conveying messages, the more often these glitches interfere with and inhibit effective communication.

LISTENING AS A COMMUNICATION TOOL

Perhaps the most important communication skill of team leaders is the ability to listen. It is also one of the rarest. Are you a good listener? Consider the following questions:

1. When in a group of people, are you more likely to talk or to listen?
2. When talking with someone, do you often interrupt the speaker before he completes the statement?
3. When talking with someone, do you find yourself tuning out and thinking ahead to your response?
4. When talking with someone, would you be able to paraphrase what the speaker has said and repeat it?
5. In conversation, do you tend to state your opinion before other speakers have made their case?
6. When talking with someone, do you give her your full attention or do you continue with other tasks simultaneously?
7. When you do not understand, do you ask for clarification?
8. In meetings, do you tend to daydream or stray from the subject?
9. When talking with someone, do you fidget and sneak glances at your watch?
10. Do you ever finish statements for people who do not move the conversation along fast enough?

A skilled listener will respond to these questions as follows:

1. Listen	6. Full attention
2. No	7. Yes
3. No	8. No
4. Yes	9. No
5. No	10. No

A better way to find out if you are a good listener is to ask a friend, your spouse, a fellow team leader, or an employee you can trust to give you an objective answer. Do not be overly concerned if you find you are a poor listener. Listening is a skill, and like all skills, it can be developed. To become a good listener, you need to know (1) what listening is, (2) the barriers that inhibit listening, and (3) the strategies that promote effective listening.

TEAMWORK TIP

"The opposite of talking isn't listening. The opposite of talking is waiting."

—Fran Leibowitz, American writer

What Is Listening?

Hearing is a natural process, but listening is not. A person with highly sensitive hearing abilities can be a poor listener. Conversely, a person with impaired hearing can be an excellent listener. Hearing is the physiological process of decoding sound waves. Listening requires perception. In this book, we define listening as follows:

> ***Listening*** *is receiving a message, correctly decoding it, and accurately perceiving what is meant.*

Inhibitors of Effective Listening

Listening breaks down when the receiver does not accurately perceive a message. Several inhibitors can cause this to happen:

- Lack of concentration
- Preconceived notions
- Thinking ahead
- Interruptions
- Tuning out

To perceive a message accurately, listeners must *concentrate on what is being said, and how it is being said.* Another part of effective listening is properly reading nonverbal cues (covered in the next section).

Concentration requires the listener to eliminate as many extraneous distractions as possible and to mentally shut out the rest. *Preconceived notions* can cause team leaders to make premature judgements that turn out to be wrong. Be patient, wait, and listen.

Team leaders who jump ahead to where they think the conversation is going may get there and find they are all alone. *Thinking ahead* is typically a response to being hurried, but it takes less time to hear an employee out than it does to start over after jumping ahead in the wrong direction.

Interruptions not only inhibit effective listening—they also frustrate and often confuse the speaker. If clarification is needed during a conversation, make a mental note of it and wait for the speaker to reach a stopping point. (Mental notes are preferable to written notes. The act of writing can distract the speaker or cause the listener to miss a critical point. If you find it necessary to make written notes, keep them short.)

Tuning out also inhibits effective listening. Some people become skilled at using body language that makes it appear they are listening when their mind is actually focusing elsewhere. Team leaders should avoid the temptation to engage in such ploys. A skilled speaker may ask you to repeat what he or she just said.

Strategies for Effective Listening

- ✓ Remove all distractions.
- ✓ Put the speaker at ease.
- ✓ Look directly at the speaker.
- ✓ Concentrate on what is being said.
- ✓ Watch for nonverbal cues.
- ✓ Make note of the speaker's tone.
- ✓ Be patient and wait.
- ✓ Ask clarifying questions.
- ✓ Paraphrase and repeat what the speaker has said.
- ✓ Control your emotions.

FIGURE 5.3 Strategies team leaders can use to improve their listening skills.

Team leaders can become effective listeners by applying the simple strategies listed in Figure 5.3.

NONVERBAL COMMUNICATION

Nonverbal messages represent one of the least understood but most powerful modes of communication. Nonverbal messages are often more telling than verbal ones, provided the receiver is attentive and able to read them.

Nonverbal communication is popularly called "body language." *Body* language is only part of nonverbal communication, however. There are actually three components: body factors, voice factors, and proximity factors (see Figure 5.4).

Body Factors

A person's posture, poses, facial expressions, gestures, and dress—body factors—can convey a message. Even such extras as makeup or lack of it, well-groomed or unkempt hair, and shined or scruffy shoes can convey a message. Team leaders should be attentive to these body factors and how they add to or detract from their and their team members' verbal messages.

One of the keys to understanding nonverbal cues lies in the concept of congruence. Are the spoken message and the nonverbal message consistent

Components of Nonverbal Communication

Body Factors

- Posture
- Dress
- Gestures
- Facial expressions
- Poses

Voice Factors

- Volume
- Pitch
- Tone
- Rate of speech
- Steady or shaky

Proximity Factors

- Relative positions
- Physical arrangements
- Color
- Texture

FIGURE 5.4 There are many elements of non-verbal communication.

with each other? They should be. To illustrate this point, consider the hypothetical example of Chem-Tech Company. An important element of the company's corporate culture is attractive, conservative dress. This is especially important for Chem-Tech's sales force. For the men, white shirts, dark suits, and shined shoes are the norm.

John McIntire is an effective sales representative, but lately he has taken to flashy dressing. He wears loud sport coats, open-neck print shirts, and casual shoes. When questioned by his team leader, John said he understands the dress code and agrees with it. He is demonstrating incongruence. His verbal message says one thing, but his nonverbal message says another. This is an exaggerated example; incongruence is not always so obvious. A simple facial expression or a subtle gesture can be an indicator of incongruence.

When verbal and nonverbal messages are not congruent, team leaders should dig a little deeper. An effective way to deal with incongruence is to gently, but frankly, confront it. A simple statement such as, "Cindy, your words agree with me, but your eyes disagree," can help draw an employee out so the team leader gets the real message.

Voice Factors

Voice factors also figure into nonverbal communication. In addition to listening to employees' words, team leaders should listen for voice factors such as volume, tone, pitch, and rate of speech. These factors can reveal feelings of anger, fear, impatience, uncertainty, interest, acceptance, confidence, and so on.

As with body factors, it is important to look for congruence. It is also advisable to look for groups of nonverbal cues. A single cue taken out of context has little meaning. But as one of a group of cues, it can take on significance. For example, if you look through an office window and see a man leaning over a desk and pounding his fist on it, it would be tempting to interpret this as a gesture of anger. But what kind of look does he have on his face? Is his facial expression congruent with desk-pounding anger—or could he simply be trying to knock loose a desk drawer that is stuck? On the other hand, if you saw him pounding the desk with a frown on his face and heard him yelling in an agitated tone, your assumption of anger would be well based. He might be angry just because his desk drawer is stuck, but nonetheless he would be angry.

Proximity Factors

Proximity factors range from where you position yourself when talking with an employee to how your office is arranged, the color of the walls, and the types of fixtures and decorations. A team leader who sits next to an employee conveys a different message than one who sits across a desk from the employee. A team leader who makes his or her office a comfortable place to visit is sending a message that invites communication. A team leader who maintains a stark, impersonal office sends the opposite message. To send the nonverbal message that employees are welcome to stop by and talk, consider the following suggestions:

- Have comfortable chairs available for visitors.
- Arrange chairs so you can sit beside visitors rather than behind the desk.
- Choose soft, soothing colors rather than harsh, stark, or overly bright or busy colors.
- If possible, have refreshments such as coffee, soda, and snacks available for visitors.

VERBAL COMMUNICATION

For team leaders, verbal communication ranks close in importance to listening. Team leaders can improve their verbal communication skills by being attentive to the following factors.

- *Interest.* When speaking with employees, show an interest in the topic. Show that you are sincerely interested in communicating your message to them. Demonstrate interest in the receivers of the message, as well. Look them in the eye, or if in a group, spread your eye contact evenly among all receivers. If you sound bored, reluctant, or indifferent, employees will notice and follow your example.
- *Attitude.* A positive, friendly attitude enhances verbal communication. A caustic, superior, condescending, disinterested, or argumentative attitude will shut off communication. Be patient, be friendly, and smile.
- *Flexibility.* Be flexible. For example, if you call your team members together to explain a new company policy but find they are uniformly focused on a problem that is disrupting their work schedule, be flexible enough to put your message aside for now and deal with the current problem. Until the employees work through what's on their minds, they will not be good listeners.
- *Tact.* Tact is an important ingredient in verbal communication, particularly when delivering a sensitive or potentially controversial message. Tact has been referred to as the ability to hammer in the nail without breaking the board. The key to tactful verbal communication is thinking before talking.
- *Courtesy.* Being courteous means showing appropriate concern for the needs of the receiver. Calling a team meeting ten minutes before quitting time, for example, is discourteous and will inhibit communication. Courtesy also dictates that the team leader not monopolize. When communicating verbally, give the receiver ample opportunities to seek clarification and to state her point of view.

COMMUNICATING CORRECTIVE FEEDBACK

In dealing with any group of employees, it is inevitable that team leaders will need to give corrective feedback. This information helps them improve their performance. To be effective, however, corrective feedback must be communicated properly. The following guidelines enhance the effectiveness of corrective feedback:[1]

- *Be positive.* To be corrective, feedback must be accepted and acted on by the employee. This is most likely to happen if it is delivered in a positive manner. Give the employee the necessary corrective feedback, but don't focus only on the negative. Find something positive to say.

- *Be prepared.* Focus on facts. Do not discuss personality traits. Give specific examples of the behavior you would like to see corrected.
- *Be realistic.* Make sure the behaviors you want to change are within the control of the employee. Don't expect an employee to correct a behavior he or she does not control. Tell the employee about his behavior, ask for his input, and listen carefully when that input is given.

IMPROVING COMMUNICATION IN TEAMS

Effective communication is a must for team leaders. The following strategies can be used to improve your communication skills:[2]

- *Keep up to date.* Stay up to date with information that might affect your team. You cannot communicate what you don't know.
- *Prioritize, and determine time constraints.* Communicating does not mean passing on to employees everything you learn. Such an approach creates overload and inhibits communication. Analyze the information you receive and decide which of it your employees need to know. Then prioritize it from *urgent* to *when time permits,* and share the information accordingly.
- *Decide whom to inform.* Once you prioritize your information, decide who needs to know it. Employees have enough to keep up with without receiving information they don't need. Correspondingly, don't withhold information that employees do need. Achieving this balance improves communication.
- *Determine how to communicate.* There are a variety of ways to communicate (*e.g.,* verbally, electronically, one-on-one, in groups). A combination of methods is usually the most effective. The next section deals with this strategy in more depth.
- *Communicate the information.* Telling your employees what you want them to know is just one step in effective communication. Tell, ask, listen, paraphrase, and follow up. Ask questions to determine if your message has been understood. Encourage employees to ask clarifying questions. Agree on the next steps (*i.e.,* what they should do with the information).
- *Check accuracy and get feedback.* Check to see that your communication was received accurately. Can employees paraphrase and repeat your message? Are employees undertaking the next steps as agreed? Solicit feedback from employees to ensure that their understanding has not changed and that progress is being made.

Selecting the Appropriate Communication Method[3]

One of the steps recommended in the previous section was "determine how to communicate." Because most workplace communication is either verbal or written, team leaders need to know how to make the most effective use of each. The following lists summarize the situations in which written and verbal communication should and should not be used.

Written communication is *least* effective in the following instances:

- *When communicating a message requiring immediate action on the part of employees.* The more appropriate approach in such a case is to communicate the message verbally and then follow it up in writing.
- *When commending an employee for doing a good job.* This should be done verbally and publicly, then followed up in writing.
- *When reprimanding an employee for poor performance.* This message can be communicated most effectively verbally and in private. This is particularly true for occasional offenses. Reprimands should be followed up in writing if they concern serious problems.
- *When resolving conflict among employees about work-related problems.* The necessary communication in such instances is most effectively conducted verbally and in private.

Verbal communication is *least* effective in the following instances:

- *When communicating a message requiring future action on the part of employees.* Such messages are most effectively communicated in writing.
- *When communicating general information such as company policies, personnel information, directives, or orders.*
- *When communicating work progress to an immediate supervisor or a higher manager.*
- *When promoting a safety campaign.*

By choosing the method of communication appropriate to the situation at hand, team leaders can enhance the effectiveness of their communication.

Summary

1. Communication is the oil that lubricates the machinery of human interaction in teams. With sufficient training and practice, people—regardless of their innate capabilities—can learn to communicate effectively.

2. Communication is the transfer of information that is received and fully understood from one source to another. Effective communication occurs when, in addition to being received and understood, the information is acted on in the desired manner.

3. Communication occurs on various levels, including one-on-one, team, organization, and community.

4. Communication as a process involves several components: the sender, the receiver, the medium, and the message. The sender originates the message. The receiver is the person or group for whom the message is intended. The message is the information conveyed, and the medium is the vehicle used to convey the message.

5. Common inhibitors of communication include differences in meaning, insufficient trust, information overload, interference, condescending tones, listening problems, premature judgements, inaccurate assumptions, and technological glitches.

6. Listening involves receiving a message, correctly decoding it, and accurately perceiving what it means. Common inhibitors of listening include lack of concentration, preconceived ideas, thinking ahead, interruptions, and tuning out.

7. Nonverbal messages are one of the least understood but most powerful modes of communication. There are three components of nonverbal communication: body factors, voice factors, and proximity factors.

8. Verbal communication can be improved by being attentive to interest, attitude, flexibility, tact, and courtesy.

9. When communicating corrective feedback to employees, team leaders should be positive, prepared, and realistic.

10. Team leaders can improve communication in their teams by keeping up to date; prioritizing, and determining time constraints; deciding whom to inform; and determining how best to communicate the message. When communicating a message, team leaders should tell, ask, listen, paraphrase, check accuracy, and get feedback.

Key Terms and Concepts

Body factors
Communication
Community-level communication
Congruence
Corrective feedback
Differences in meaning
Effective communication
Inaccurate assumptions
Incongruence
Information overload
Insufficient trust
Interference
Listening
Medium
Message
Nonverbal communication
One-on-one level communication
Organization-level communication

Preconceived ideas
Proximity factors
Receiver
Sender
Tact
Team- or unit-level communication
Voice factors
Verbal communication
Written communication

Review Questions

1. Define the term "communication."
2. What is the difference between communication and effective communication?
3. Explain the communication process.
4. What are the inhibitors of communication (list five)?
5. Define the term "listening."
6. What are the inhibitors of effective listening (list five)?
7. What are the three components of nonverbal communication?
8. Explain the concept of congruence.
9. How can team leaders effectively communicate corrective feedback?

EFFECTIVE TEAMWORK SIMULATION CASES

The following cases deal with specific issues relating to the implementation of effective teamwork. Each case represents a meeting of Marcee McPhee and Pete Fared, engineers and team leaders at Mac-Tech, Inc., a technology firm with 526 employees. McPhee is the leader of Team A, and Fared leads Team B. McPhee and Fared are not just colleagues; they are friends, and their friendship goes all the way back to college. They both attended the same engineering school and graduated in the same class. Once a week they meet for lunch and discuss problems, progress, issues, and concerns. These cases chronicle their luncheon conversations and invite the reader to discuss the issues Fared and McPhee deal with.

CASE 5.1 Inhibitors of Effective Communication

"I've been reading that handbook on effective communication that the boss gave all the team leaders," commented Pete Fared. "At least now I understand why communication seems to be so difficult in my team." "Me too," said Marcee McPhee. "I used to think I could just tell my team members what they needed to know and everyone would understand. In fact, it bugged me a little when they didn't." Fared and McPhee were working on improving communication in their teams, the most recent step in their company's implementation of teamwork.

"No offense, but it's the female members of my team I have the most difficulty communicating with," admitted Fared. "Except for one or two of them, I can communicate pretty well with the men in my team. But the women and I usually seem to be on different planets." "The communication problems on my team don't seem to be gender-based," said McPhee. "But I do have problems communicating with those three technicians from Russia. They speak English better than I do, but we just don't seem to be on the same wavelength most of the time."

Discussion Questions

1. Have you ever been in a situation where you needed to communicate with someone but found it difficult to do so? What were the inhibitors that caused the difficulty?
2. Examine the problems Fared and McPhee are having with communication in their teams. What inhibitors might be causing their problems?

CASE 5.2 Listening Problems in the Team

"Do your team members listen to you?" asked Marcee McPhee. Looking across the table, Pete Fared could tell that something was bothering his colleague. "I don't know," he replied. "I mean, I guess so. Why do you ask?" "Because I have a couple of team members who are really poor listeners. Every time I give instructions, they claim to have heard me, but within the hour they make a mistake that shows me they didn't." "That happens in my team sometimes too," acknowledged Fared. "In fact, it happened just yesterday."

Fared told McPhee about a situation that had occurred in his team only the day before. He had called a team meeting to discuss a new procedure handed down by management that was to be put in place immediately. After he had explained the new procedure and why it was important, several team members had asked questions. In fact, everyone on the team had asked at least one question except one team member. This particular employee had gone back to work and proceeded to pick up right where he had left off using the old procedure. Fortunately, another team member noticed and stopped him before too much damage had been done.

Discussion Questions

1. Do you know someone who is a poor listener? What kinds of problems are caused by this individual's poor listening skills?
2. Put yourself in Pete Fared's place. In the future, how would you make sure that the problem employee had heard instructions, understood them, accepted them, and was prepared to act properly on them?

CASE 5.3 Problems Communicating Corrective Feedback

"Marcee, the worst part of my job is giving corrective feedback to my team members. I'm not good at it, and I don't like doing it," said a frustrated Pete Fared. Fared had given corrective feedback to one of his team members that very morning, and the discussion had not gone well. He explained to Marcee McPhee that the team member in question had been defensive, had denied most of what Fared had said, and had left the meeting upset. "I've got to get better at this," said Fared with a sigh.

"I don't like correcting team members either," admitted McPhee. "But it has to be done. I've gotten better at it over the years, but when I first became a team leader, I would actually get sick to my stomach in anticipation of a corrective-feedback meeting with an employee."

"Well, at least I'm not the only one who has trouble with it," said Fared. "How can I get better at giving corrective feedback, Marcee?"

Discussion Questions

1. Giving corrective feedback is difficult for some people. How do you feel about it? Are you comfortable telling someone they have performed poorly or done something wrong?
2. Put yourself in Marcee McPhee's place. What advice would you give Pete Fared?

Endnotes

1. Robert A. Luke, "How to Give Corrective Feedback to Employees," *Supervising Management*, March 1980, p. 7.
2. Kim McKinnon, "Six Steps to Improved Communication," *Supervising Management*, February 1999, p. 9.
3. D. A. Level, "Communication and Situation," *Journal of Business Communication*, 9 (1992), pp. 19–25.

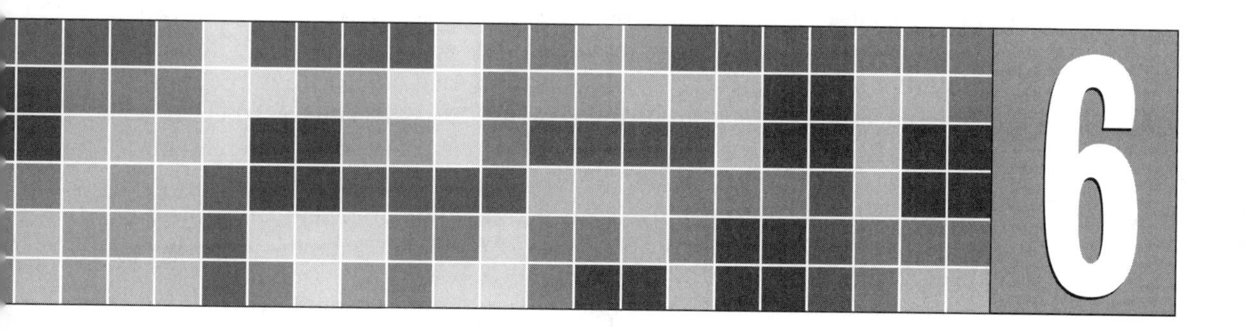

Develop Conflict-Management Skills

The only way to get the best of an argument is to avoid it.

—Dale Carnegie

OBJECTIVES

- Name and explain the causes of conflict in teams.
- List and define the various ways people react to conflict.
- Discuss the reasons conflict-management skills are important.
- Explain the steps team leaders can take in resolving conflict.
- Explain when conflict should be stimulated.
- Explain the concept of communication in conflict resolution.
- Explain how to handle angry team members.
- Explain how to overcome territorial behavior in teams.
- Explain how to overcome negativity in teams.

Human conflict is normal in a highly competitive and often stressful workplace. One of the human-relations skills necessary to the success of teams is the ability of team members to disagree with each other without being disagreeable. Even if most team members have this skill, however, there is no guarantee that conflicts will not arise. When people work together, no matter how committed they are to a common goal, conflict is likely to occur. Consequently, all team members must come to be proficient in managing conflict. This chapter explains the ways people can be catalysts in resolving conflict within teams.

CAUSES OF CONFLICT IN TEAMS

The most common causes of conflict in teams are predictable. They are shown in Figure 6.1.

> ***Conflict** occurs when a person's desires are frustrated or needs are threatened by another person.*

Limited resources can lead to conflict because it is not uncommon for teams to have fewer resources (e.g., funds, supplies, personnel, time, equipment) than are needed to complete a job. When this happens, who gets the resources and in what amounts? *Incompatible goals*, such as conflicts between personal needs and the needs of the team, are inevitable in the workplace. *Role ambiguity* can result in blurred "turf lines." This makes it difficult to know who is responsible and who has authority. *Different values*, such as job security and risk taking, create the potential for conflict. *Different perceptions* of situations, based on peoples' differing backgrounds,

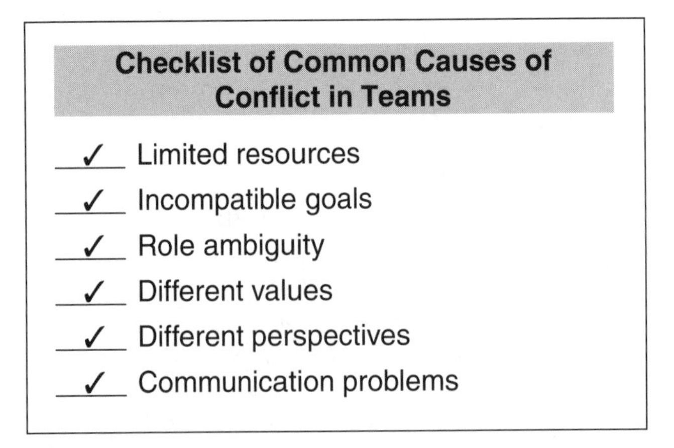

FIGURE 6.1 Conflict in teams stems from many potential causes.

values, beliefs, and individual circumstances, are common—particularly in an increasingly diverse workplace. The final predictable cause of conflict is *poor communication*. Because communication is never perfect, communication-based conflict should be expected. Improving the communication skills of all team members should be an on-going goal of team leaders.

HOW PEOPLE REACT TO CONFLICT

To deal with conflict effectively, team leaders need to understand the various ways people react to conflict. One typical reaction is *competition*, in which one person attempts to win while causing the other to lose. The opposite reaction is *accommodation*, in which one person puts the needs of the other first, and lets her win. *Compromise* is a reaction in which two opposing parties attempt to work out a solution that helps both to the greatest extent. *Collaboration* involves both parties working together to find an acceptable solution for both. *Avoidance* involves shrinking away from conflict, a common reaction of people who are uncomfortable facing conflict and dealing with it.

In some situations, one type of reaction to conflict can be more appropriate than another. Team leaders who are responsible for resolving conflict need to understand these distinctions.

- *Competing* is appropriate when quick action is vital or when important but potentially unpopular actions must be taken.
- *Collaborating* is appropriate when it is important to work through feelings that are interfering with interpersonal relationships.
- *Avoiding* is appropriate when you perceive no chance of satisfying your concerns or when you want to cool down and have time to regain a positive perspective.
- *Accommodating* is appropriate when you are outmatched and losing anyway, or when harmony and stability are more important than the issue at hand.

WHY CONFLICT-MANAGEMENT SKILLS ARE IMPORTANT

"Conflict management" is a broad term that encompasses conflict resolution, conflict stimulation, and the appropriate use of each. Regardless of whether team members accommodate, compromise, compete, collaborate, or avoid, they all become stressed over conflict. In fact, conflict is a major cause of workplace stress, which is why it is important that team leaders be skilled at managing conflict. Workplace-related stress can have an adverse affect in many ways:

- *Employee performance.* Employees who become overly stressed do not perform at peak levels in terms of either quality or productivity.
- *Customer service/satisfaction.* Employees who are overly stressed are unable to maintain the positive, helpful attitude needed to properly serve and satisfy customers.
- *Employee safety.* Employees who are overly stressed are more accident prone and as a result can be dangerous to other employees.
- *Employee health.* Overly stressed employees soon become sick employees. Stress can increase blood pressure and heart rate and produce gastrointestinal problems, such as ulcers.
- *Absenteeism and tardiness.* Overly stressed employees tend to be tardy and absent more than they would be otherwise. Their absenteeism and tardiness can have a detrimental effect on the team's overall performance.

TEAMWORK TIP

"Nature has given to men one tongue, but two ears, that we may hear from others twice as much as we speak."

—Epictetus, Greek philosopher

HOW TEAM LEADERS SHOULD HANDLE CONFLICT

When conflict arises, team leaders need to resolve it in ways that serve the team's long-term best interests. This will keep the conflict from becoming a detriment to performance.

The following sequence of steps can be used by team leaders attempting to resolve conflict:

1. Determine the importance of the issue to all people involved.
2. Determine whether the people involved are willing and able to discuss the issue in a positive manner.
3. Select a private place where the issue can be discussed confidentially by everyone involved.
4. Make sure that both sides understand they are responsible for both the problem and the solution.
5. Solicit opening comments from both sides. Let them express their concerns, feelings, ideas, and thoughts—in a nonaccusatory manner.
6. Guide participants toward a clear and specific definition of the problem.
7. Encourage participants to propose solutions, while you listen carefully. Examine the problem from a variety of perspectives, and discuss all solutions proposed.

8. Evaluate the costs versus the gains (cost-benefit analysis) of all proposed solutions and discuss them openly. Choose the best solution.
9. Reflect on the issue and discuss the conflict-resolution process. Encourage participants to express their opinions about how the process might be improved.

WHEN CONFLICT SHOULD BE STIMULATED

It is possible for teams to have too little conflict. This can occur when employees become overly comfortable or when management has effectively suppressed critical thinking, innovation, and creativity. When this occurs, stagnation generally results. Stagnant teams need to be shaken up or they will die. Team leaders can do this by stimulating positive conflict aimed at revitalizing the organization. A "yes" response to any of the following questions suggests a need for conflict stimulation:

- Do team members always agree with you and tell you only what you want to hear?
- Are team members afraid to admit they need help or that they've made mistakes?
- Do team members focus more on reaching an agreement than on arriving at the best decision?
- Do team members focus more on getting along with others than on accomplishing objectives?
- Do team members place more emphasis on not hurting feelings than on arriving at quality decisions?
- Do team members place more emphasis on being popular than on job performance?
- Are team members highly resistant to change?
- Do team members avoid proposing new ideas?

It may be possible to have a vital, energetic, developing, improving team without conflict, but it isn't likely. Innovation, creativity, and the change inherent in continual improvement typically breed conflict. Therefore, the complete absence of conflict can indicate the absence of vitality, and team leaders need to know how to stimulate positive conflict.

Techniques for stimulating conflict fall into three categories: *improving communication*, *involving team members*, and *changing behavior*.

- *Improving communication* helps ensure a free flow of ideas in the team. Open communication introduces a daily agitation factor that prevents stagnation and provides a mechanism for effectively dealing with the resulting conflict.

- *Involving team members* in decision making about issues that affect them helps prevent stagnation. If employees are given a voice, they will use it. The result will be positive conflict.
- *Changing behavior* may be necessary, particularly in teams that have traditionally suppressed and discouraged conflict rather than dealing with it. Team leaders in that situation may find the following procedure helpful: (a) identify the behaviors you want employees to exhibit; (b) communicate with employees about those behaviors, so they understand what is expected; (c) reinforce the desired behaviors; and (d) handle conflict as it emerges using the procedures set forth in the previous section.

COMMUNICATION IN CONFLICT RESOLUTION

The point was made in the previous section that human conflict in teams is normal, to be expected, and, in certain instances, to be promoted. Managing conflict in essence means resolving conflict when it has negative effects and promoting conflict when it helps avoid stagnation. In both senses, communication is critical.

The following guidelines can be used to improve communication in managing conflict:

- *The initial attitude of those involved in the conflict can predetermine the outcome.* If a person enters into a situation spoiling for a fight, he will probably get one. Communicating with and convincing either or both parties to view the conflict as an opportunity to cooperatively solve a problem can help predetermine a positive outcome.
- *When possible, conflict guidelines should be in place before conflicts occur.* It is not uncommon that conflict is exacerbated by disagreements over how it should be resolved. Before entering into a potentially conflictual situation, make sure all parties understand how decisions will be made, who has the right to give input, and what issues are relevant.
- *Assigning blame should not be allowed.* People in conflict situations often blame each other. If human interaction is allowed to get hung up on assigning blame, it will never move forward. The approach that says, "We have a problem. How can we work together to solve it?" is more likely to result in a positive solution than arguing over who is to blame.
- *"More of the same" solutions should be eliminated.* When a particular strategy for resolving conflict has been tried but has not been effective, it is not productive to continue using it. Some team leaders get stuck on a particular approach and stay with it even if it clearly doesn't work. Try something new rather than using "more of the same" solutions.
- *Maintain trust by keeping promises.* Trust is fundamental to all aspects of teamwork and is especially important when managing conflict. Conflict

cannot be effectively managed by someone who is not trusted. Consequently, team leaders must keep their promises and, in so doing, build trust among employees.

HANDLING ANGRY TEAM MEMBERS

Problems are never solved when employees or their team leaders are in a state of anger. Worse yet, angry employees can become violent if not handled properly. Therefore, it is critical that team leaders know how to deal with angry team members. The strategies in this section will help team leaders defuse angry situations before they erupt into even worse problems.

Behaviors to Avoid

Team leaders may feel compelled to take immediate disciplinary action against an angry employee. Although such action might be warranted, and although it may need to be taken at a later point, calming the angry employee should be the first priority. Team leaders must avoid the following behaviors when dealing with angry employees:

- *Becoming angry and responding in kind.* Responding in kind to an angry employee is like pouring gasoline on a fire: things are just going to get worse, and sooner rather than later. Responding in kind to anger is not an option for team leaders.
- *Walking away or hanging up the telephone.* When dealing with an angry person, it is always tempting to think, "I don't have to put up with this," and to walk away or hang up the telephone. Although this might avoid the negativity temporarily, it will probably make matters worse in the long run. It is better to deal with a negative situation immediately, before it escalates.
- *Pointing out that the employee is being rude.* Calling an angry employee "rude" most often will make that employee even angrier. In a confrontational situation, few things sting worse than the obvious truth. This does not mean that the team leader should ignore the anger. Quite the contrary. The anger should be acknowledged, but not by pointing out how rude it is. There is more about acknowledging anger later in this chapter.

What Team Leaders Should Do

The two most important things to remember when dealing with angry employees are: (1) remain calm and (2) focus on the problem, not the anger. Of course this approach is more easily said than done. The following strategies will help team leaders stay calm and focused when dealing with an angry employee:

- *Control your breathing.* In stressful situations, particularly those involving anger, breathing tends to become rapid and irregular, which can make you feel out of control. If this happens, take several deep breaths and relax. This will help regulate your breathing and settle your nerves.
- *Look through the anger for the real message.* People who are upset to the point of being angry often say things they don't really mean and express themselves in broad generalities, implying that everything and everyone is wrong. Team leaders who take such attacks personally are prone to defensiveness or the urge to fight back. Suppress this natural urge, and try to see through the diatribe to what is really bothering the employee.
- *Be aware of your tone of voice and body language.* If approached with a condescending, argumentative, or frustrated tone of voice, an angry employee can become even angrier. Negative body language such as rolling of the eyes, unbelieving facial expressions, or confrontational stances can also exacerbate the situation. Sit up straight, look the employee in the eye, and show that you are listening.

How to Calm an Angry Employee

The previous section provided strategies to help team leaders stay calm when dealing with angry employees. Although staying calm is critical, it's just the first step. The next step is to calm the angry employee. The following strategies can help in this regard, so that the problem at the heart of the employee's anger can be dealt with:

- *Do not interrupt or disagree with the angry employee.* Sometimes the best thing to do is simply to allow the employee to vent. Find a private place, let the employee speak her mind, and just listen. Do not interrupt or disagree. There will be plenty of time to get the facts straight after the anger has subsided. If the employee begins to ramble and repeat herself, you can lead her back to the point in question by waiting for a convenient opening and then saying, "Excuse me. I understand your first two points, but I need you to clarify the third point for me." The employee may not actually have made any points, but that doesn't matter. This technique will usually help her get focused. Often, after letting an employee vent, team leaders find that no further action is necessary. All the employee needed was for someone to listen.
- *Paraphrase what the employee says and repeat it back.* This lets the employee know that you are listening to what he is saying. Nothing aggravates the frustration of an angry employee more than not being listened to. In fact, the source of many employees' anger is their feeling that no one will listen to them. Another advantage of this strategy is verification. If what you paraphrase and repeat is not what the employee had meant to communicate, he can tell you immediately. Then you can ask him to clarify your understanding.

■ *Acknowledge the anger.* Simply acknowledging the employee's anger goes a long way toward calming her down. Just say, "I can see you are really angry about this." This has the effect of holding up a mirror and letting the employee watch herself be angry. If we all looked in a mirror when acting out our anger, there would be fewer angry displays. In addition, acknowledgment shows understanding.

■ *Encourage the angry employee to work with you in solving the problem.* This strategy does two things. First, it subtly communicates to the employee that although you will work with him, you are not going to let him simply dump the problem in your lap—something many disgruntled employees want to do. Second, it moves the employee from the complaining mode to the solution mode. You want to make this transition as quickly as possible, but without rushing the employee.

■ *Arrive at a specific solution.* In moving toward a solution, it's a good idea to ask, "What solution do you propose?" If a team leader solves a problem without involving the employee, the leader will be blamed if the solution doesn't work as planned. If the employee is part of the solution, however, she is likely to feel ownership in it and to have a corresponding incentive to ensure that it works. Before concluding a meeting with an angry employee, arrive at a specific solution, and make sure the employee understands that solution and has part ownership of it.

TEAMWORK TIP

"Anger is just one letter short of danger."

—Anonymous

OVERCOMING TERRITORIAL BEHAVIOR IN TEAMS

"Territory" in the team situation tends to be more a function of psychological boundaries than of physical ones. Often referred to as "turf protection," territoriality can manifest itself in a variety of ways.

Manifestations of Territoriality

Territorial behavior can show up in a variety of ways. The most common territorial games are explained here:[1]

■ *Occupation.* Territorial games involving occupation include actually marking office space as "mine"; playing "gatekeeper" with information; and monopolizing resources, information, access, and relationships.

■ *Information manipulation.* People who play territorial games with information subscribe to the philosophy that information is power. To feel powerful, they withhold information, bias (spin) information to suit their individual agendas, cover up information, and give out false information.

- *Intimidation.* One of the most common manifestations of territoriality is intimidation—a tactic used to frighten others away from certain turf. Intimidation can take many forms, from subtle threats to blatant aggression (physical or verbal).
- *Alliances.* Forming alliances with powerful individuals in an organization is a commonly practiced territorial game. The idea is to communicate, without having to say it, "You had better keep off my turf, or I'll get my powerful friend to cause trouble."
- *Invisible walls.* The goal of putting up invisible walls is to ensure that decisions, although already made, cannot be implemented. There are hundreds of means of creating hidden barriers, including stalling, losing paperwork, and forgetting to place an order.
- *Strategic noncompliance.* Agreeing to a decision up front, but having no intention of carrying it out, is called "strategic noncompliance." This tactic is often used to buy time to find a way to reverse the decision.
- *Discrediting.* Discrediting an individual to cast doubt on his recommendation is a common turf-protection tactic. Such an approach is called an *ad hominem* argument, meaning that if you cannot discredit the recommendation, you discredit the person making it.
- *Shunning.* Excluding an individual who threatens your turf is also a common turf-protection tactic. Shunning creates peer pressure against the excluded individual.
- *Camouflage.* This tactic can also be called "throwing up a smoke screen" or "creating fog." It involves confusing the issue by raising other distracting controversies—especially those intended to produce anxiety.
- *Filibustering.* "Filibustering" means talking a recommended action to death. It involves talking at length about concerns—usually inconsequential ones—until either the other side gives in just to stop any further discussion or the time to make the decision runs out.

Overcoming Territorial Behavior

Overcoming territorial behavior requires a two-step approach: (1) recognizing one or more of the manifestations described above and admitting that they exist and (2) creating an environment in which survival is equated with cooperation rather than territoriality. Simmons recommends the following strategies to create a cooperative environment:[2]

- *Avoid jumping to conclusions about the motives of people.* Talk to employees about territoriality versus cooperation. Ask to hear their views, and listen to what they say.
- *Attribute territorial behavior to instinct rather than to people.* Blaming people for following their natural instincts is like blaming them for eating.

The better approach is to show them that their survival instinct is tied to cooperation, not territoriality. This is done by rewarding cooperation and applying negative reinforcement to territorial behavior.

- *Ensure that no employee feels attacked.* Remember that the survival instinct drives territorial behavior. Attacking employees, or even leading them to feel attacked, triggers their survival instinct. To change territorial behavior, it is necessary to put employees at ease.

- *Avoid generalizations.* When an employee exhibits territorial behavior, deal with the specific situation as opposed to generalizing it. It is a mistake to witness territorial behavior on the part of one employee and to respond to it on a group level. Deal with the individual who exhibits the behavior, and deal in specifics.

- *Understand irrational fears.* The survival instinct is a powerful motivator that can cause employees to cling to irrational fears. Team leaders should consider this point when dealing with employees who find it difficult to let go of survival behaviors. Be firm but patient, and never deal with an employee's fears in a denigrating or condescending manner.

- *Respect each individual's perspective.* In a way, an individual's perspective or opinion is part of his psychological territory. Failure to respect employees' perspectives is the same as threatening their territory. When challenging territorial behavior, let employees explain their perspectives, and show respect for them even if you do not agree.

- *Consider the employee's point of view.* In addition to demonstrating respect for employees' perspectives, team leaders should try to step into their shoes. How would you feel if you were the employee? Sensitivity to and patience with the employee's point of view are critical when trying to overcome territorial behavior.

OVERCOMING NEGATIVITY IN TEAMS

Negativity is any behavior on the part of any member that works against the optimum performance of the team. The motivation behind negativity can be as different and varied as the employees who manifest it. Negative behavior can be categorized as follows:

- Control disputes (Who is in charge?)
- Dependence/independence issues (I need/I don't need . . .)
- Need for attention/responsibility (Give me strokes/Put me in charge.)
- Authority issues (I am in control.)
- Loyalty issues (Support me no matter what.)

Recognizing Negativity

Team leaders should be alert to signs of negativity in the workplace, because negativity is contagious. It can spread quickly throughout a team, dampening morale and inhibiting performance. What follows are indicators of the negativity syndrome that team leaders should watch for:

- *"I can't" attitudes.* Employees in a team committed to continuous improvement must have can-do attitudes. If "I can't" is heard regularly, negativity has crept into the team.
- *"They" mentality.* In high-performance teams, employees use "we" when talking about the team. If employees refer to their team as "they," negativity has gained a foothold.
- *Critical conversation.* In high-performance teams, coffee-break conversation is about positive, work-related topics or topics of personal interest. When conversation is typically critical, negative, and judgmental, negativity has set in. Some team leaders subscribe to the philosophy that employees are not happy unless they are complaining; this is a dangerous attitude. Positive, improvement-oriented employees complain to their team leaders about conditions that inhibit performance, but they don't criticize and whine during coffee breaks.
- *Blame fixing.* In a high-performance team, employees fix problems, not blame. If blame fixing and finger pointing are common in a team, negativity is at work.

Overcoming Negativity

When team leaders identify negativity in their teams, they should take appropriate steps to eliminate it. The following strategies can be used to overcome negativity in teams:

- *Communicate.* Frequent, ongoing, effective communication is the best defense against negativity in teams as well as the best offense for overcoming negativity that has already set in. Team communication can be made more effective by (1) acknowledging innovation, suggestions, and concerns; (2) sharing information so that all employees are informed; (3) encouraging open, frank discussion during meetings; (4) celebrating milestones; (5) giving employees ownership of their jobs; and (6) promoting teamwork.
- *Establish clear expectations.* Make sure all employees know what is expected of them as individuals and as members of the team. They also need to know how and to whom they are accountable for what is expected.
- *Provide for anxiety venting.* The workplace can be stressful in even the best organizations and at the best of times. Deadlines, performance standards, budget pressures, and competition can all produce anxiety in employees. Consequently, team leaders need to give their direct reports opportunities to

vent in a non-threatening, affirming environment. This type of environment is characterized by a team leader who listens supportively and does not shoot the messenger. The team leader listens without interrupting, thinking ahead, focusing on preconceived ideas, or tuning out.

- *Build trust.* Negativity cannot flourish in an atmosphere of trust. Team leaders build trust between themselves and employees and among employees by applying the following strategies: (1) always deliver what is promised; (2) remain open to suggestions; (3) take an interest in the development and welfare of employees; (4) be tactfully honest with employees at all times; (5) lend a hand when necessary; (6) accept blame, but share credit; (7) maintain a steady, pleasant temperament even when under stress; and (8) make sure that criticism is constructive and delivered in an affirming way.
- *Involve employees.* It's hard to criticize the way things are done when you participate in determining how they are done. Involve employees by asking their opinions, soliciting their feedback, and making them part of the solution.

Summary

1. Conflict arises when a person's desires are frustrated or his needs are threatened by another person. Common causes of conflict in teams include: limited resources, incompatible goals, role ambiguity, different values, different perceptions, and poor communication.
2. People respond to conflict in one of the following ways: competition, accommodation, compromise, collaboration, or avoidance.
3. Conflict-management skills are important for team leaders because conflict can have a negative effect on employee performance, customer service, employee safety and health, employee attendance, and employee punctuality.
4. Team leaders should handle conflict by determining how important the issue is to all involved, ensuring that all parties are willing to discuss the issue in a positive manner, selecting a private location for the discussion, making sure that both sides understand that they are responsible for finding solutions, allowing both sides to have their say without interruption, guiding participants to a clear definition of the problem, and encouraging participants to propose solutions.
5. Conflict should be stimulated when a team appears to be stagnant. Conflict can result in creative ideas and innovative approaches if handled properly.
6. Communication guidelines for conflict resolution include the following: the initial attitude of those involved can predetermine the outcome, conflict guidelines should be in place before attempting to resolve an issue, assigning blame should not be allowed, "more of the same" solutions should not be allowed, and trust should be maintained by keeping promises.

7. When handling angry team members, it is important that team leaders avoid the following behaviors: becoming angry and responding in kind, walking away or hanging up the telephone, and pointing out that the angry person is being rude.

8. The following behaviors are helpful in dealing with an angry team member: control your breathing (take deep breaths), look through the anger for the real message, and be aware of your voice tone and body language.

9. The following strategies can help calm an angry team member: avoid interrupting or contradicting, paraphrase and repeat back what has been said, acknowledge the anger, encourage the team member to work with you in solving the problem, and arrive at a specific solution.

10. Territorial behavior in teams manifests itself in the following ways: occupation, information manipulation, intimidation, alliances, invisible walls, strategic noncompliance, discrediting, shunning, camouflage, and filibustering.

11. Territorial behavior can be overcome by applying the following strategies: avoid jumping to conclusions, attribute the behavior to instinct rather than to people, ensure that no employee feels attacked, avoid generalizations, understand that people have irrational fears, respect each individual's perspective, and consider the other person's point of view.

12. Negativity in teams typically falls into the following categories: control disputes, dependence/independence issues, the need for attention/responsibility, authority issues, and loyalty issues.

Key Terms and Concepts

Accommodation
Alliances
Anxiety venting
Avoidance
Blame fixing
Camouflage
Collaboration
Competition
Compromise
Conflict management
Conflict resolution
Conflict stimulation
Critical conversation
Discredit
Filibustering
"I can't" attitudes
Information manipulation
Intimidation
Invisible wall
Negativity
Occupation
Role ambiguity
Shunning
Strategic noncompliance
Territoriality
"They" mentality

Review Questions

1. List and explain five common causes of workplace conflict.
2. Explain the various ways people respond to conflict.
3. What is the difference between *conflict resolution* and *conflict management*?
4. When should a team leader stimulate conflict?
5. How should a team leader use communication to prevent workplace conflict?
6. When dealing with angry employees, what should team leaders *not* do? What things *should* team leaders do?
7. How does a team leader best calm an angry employee?
8. List and explain five manifestations of territoriality.
9. What are five strategies team leaders can use to overcome territorial behavior?
10. What are the symptoms of negativity in a team?
11. What can team leaders do to overcome negativity in their teams?

EFFECTIVE TEAMWORK SIMULATION CASES

The following cases deal with specific issues relating to the implementation of effective teamwork. Each case represents a meeting of Marcee McPhee and Pete Fared, engineers and team leaders at Mac-Tech, Inc., a technology firm with 526 employees. McPhee is the leader of Team A, and Fared leads Team B. McPhee and Fared are not just colleagues; they are friends, and their friendship goes all the way back to college. They both attended the same engineering school and graduated in the same class. On the job, their relationship has evolved into one of mentoring, in which McPhee is helping Fared learn to be a better team leader. Once a week they meet for lunch and discuss problems, progress, issues, and concerns. These cases chronicle their luncheon conversations and invite the reader to discuss the issues Fared and McPhee deal with.

CASE 6.1 Why So Much Conflict Lately?

"Marcee, lately I feel more like a referee than a team leader," said a frustrated Pete Fared. Fared's team had been relatively conflict-free since he had taken over as the team leader. Other than a few minor disagreements between the same two team members, team conflict had been a non-issue for Fared. Lately, however, circumstances had changed. "What's different in your team?" asked McPhee. "Do you have any new team members?" Fared explained that he had three new team members but that they seemed to be good workers.

"I had a similar situation once," offered McPhee. "Two of my team members left and two new technicians joined the team. The new members had been on the job less than a week when the trouble started. At first I just took the conflict in stride, thinking my team members were all adults and would work it out." "Did they?" asked Fared. "No," conceded McPhee, sounding embarrassed. "Things just got worse."

"What did you do?" asked Fared. McPhee explained that she had called a team meeting to get to the bottom of things, but made no progress. When the team meeting led to nothing more productive than accusations and finger pointing, McPhee had changed tactics. By meeting with each team member individually, McPhee had been able to form a clearer picture of what was causing the conflict in her team. "What was the problem?" asked Fared. McPhee thought for a moment before answering. "Call it role ambiguity. Before two of my team members left, everyone on the team knew who did what. I'm not talking about formal roles here, Pete. I'm talking about the little things such as picking up supplies, making coffee, taking minutes in our team meetings, and other outside-of-the-job duties. There are a lot of little unofficial duties that have to be done to keep a team running smoothly, and my team members had always worked out who would do what according to interest, ability, seniority, and so on. It's probably that way on your team too. Good team members work these things out among themselves without even involving their team leader."

McPhee went on to explain that when the two veteran team members left, the unofficial duties distributed among team members got out of balance. Old team members expected the new team members simply to pick up where their predecessors had left off. But the new team members couldn't do that, because they didn't have the same talents and interests, nor did they understand the unwritten rules about who did what. They hadn't been on the team long enough to know.

"Major conflicts in teams can be caused by relatively minor issues," continued McPhee. "Part of the problem in my team turned out to be nothing more major than coffee." "Coffee?" exclaimed Fared, obviously having difficulty believing what he was hearing. "What's coffee got to do with it?" "Quite a lot as it turned out," answered McPhee, chuckling at the memory. She told Fared that one of the veteran team members who left had been the coffee maker for the team every morning. He always came in a little early and had the coffee steaming hot and ready when his teammates arrived. His replacement on the team, however, did not drink coffee and objected to being expected to make it. When veteran team members complained to her, the other new team member took her side in the disagreement, and before long a veterans-versus-the-new-guys mentally had set in that began to color all interactions in the team. From that point on, most of the conflicts in McPhee's team could be traced to an empty coffeepot.

"Incredible," exclaimed Fared. "I guess I'd better start interviewing my team members individually. I know the problem isn't coffee though." "How do you know?" asked McPhee. "Because I'm the coffee maker in my team, and my team members like the way I make it—strong and black." McPhee laughed at her colleague's boast and said, "If you're the coffee maker, maybe coffee is the problem." "No, it's something else," claimed Fared with surety. "I'm just not sure what." "What other things can cause conflict in teams?"

Discussion Questions

1. Have you ever been part of a team that had a lot of conflict? If so, what were the underlying causes?
2. Put yourself in Marcee McPhee's place. How would you answer Pete Fared's last question?

CASE 6.2 How Should I Handle Conflict in My Team?

"I had a little trouble in my team this morning, and I think I made a mess of things trying to fix it," said Pete Fared with a sigh. Fared had attempted to resolve an ongoing conflict between two of his team members earlier in the day, but his efforts had backfired on him. "Not only are my two team members still at odds with each other, now they are angry at me, too. I don't think I'm cut out for conflict management."

"It can't be all that bad. You're probably just overreacting," said Marcee McPhee reassuringly. "Tell me what happened."

Fared explained that two of his team members had been having disagreements off and on for about two weeks. Up until that morning, he had not attempted to intervene, hoping they would work it out themselves. But that morning in his weekly team meeting, these two employees had sniped at each other continually. Finally, when he had had enough, Fared stopped the meeting and told his feuding team members to, "grow up and start acting like adults."

Putting up her hand in the universal sign for "hold it," an incredulous McPhee said, "Wait a minute, Pete. Don't tell me you said that to these employees right in front of the whole team!" Fared nodded reluctantly. "Bad idea, huh?" "No wonder your problem employees are upset with you," said McPhee as she shook her head in disbelief.

An uncomfortable silence hung in the air until finally Fared said, "Alright, Marcee. Mea culpa. How should I have handled this situation?"

Discussion Questions

1. Pete Fared let a feud between two team members go on for two weeks hoping they would work things out. In your opinion, was this an appropriate strategy?

2. Put yourself in Marcee McPhee's place in this scenario. How would you answer Fared's last question?

CASE 6.3 How Do I Handle an Angry Employee?

"Marcee, you won't believe what happened this morning," said Pete Fared with a look that said he couldn't believe it himself. Marcee McPhee smiled at her colleague and said, "Pete, I've been around long enough and seen so much that nothing surprises me any more. But tell me anyway."

Eager to talk about it, Fared said, "I was in my office returning telephone calls when one of my team members burst through the door and yelled at me." "What did he say?" asked McPhee. "I don't remember it all word for word," admitted Fared, "but it was something like this: '*I'm sick and tired of this stupid company. None of our managers know what they are doing.*' "

"Wow!" exclaimed McPhee. "That's a pretty strong statement for an employee to make." "You're telling me it is," agreed Fared. "I felt like every manager in the building probably heard him." "What did you do?" queried McPhee. "Nothing really," admitted Fared. "I told him to lower his voice before he got both of us fired. Then I closed the door and told him to calm down. I told him to come back after lunch so we could talk about what was bothering him. That's when I called you and suggested we take an early lunch."

"Pete, I'm not sure that was the best way to handle this situation," said McPhee quietly. "I was afraid you'd say that," acknowledged Fared. "How should I have handled it?"

Discussion Questions

1. Analyze Pete Fared's response to the angry employee in this scenario. What did he do wrong? Did he do anything right?
2. Put yourself in Marcee McPhee's place. How would you answer Fared's last question?

Endnotes

1. Annette Simmons, *Territorial Games* (New York: AMACOM, American Management Association, 1998), p. 179.
2. Simmons, *Territorial Games*, p. 187.

7

Establish a Well-Defined Decision-Making Process and Empower Team Members

The best decisions come from involving the people who must carry them out.

OBJECTIVES

- Define "decision making."
- Explain how problems are associated with decision making.
- Outline the decision-making process for teams.
- Describe the two categories of decision-making models for teams.
- Explain the concept of empowering team members in decision making.
- Explain the role of information in decision making.
- Discuss the importance of creativity in decision making.

Decision making and problem solving are important responsibilities of team leaders, who make decisions and solve problems within the specified limits of their range of authority. These limits should be clearly understood so there is no question as to which decisions team leaders are allowed to make, which problems they have the authority to solve, and the extent to which team members should be involved in these processes.

DECISION MAKING DEFINED

People make decisions every day. Some are minor (What should I wear to work today? What should I have for breakfast?); some are major (Should I accept a job offer in another city? Should I buy a new house?). Regardless of the nature of the decision, "decision making" can be defined as follows:

> ***Decision making** is the process of choosing one alternative from among two or more alternatives. (Ideally, the alternative chosen is the best or at least the optimum alternative available.)*

Decisions can be compared to fuel in an engine. Decision making keeps the engine (team) running. In typical cases, work cannot progress until a decision is made.

Consider the following example: Because the computer network is down, XYZ Company has fallen behind schedule. The company cannot complete an important contract on time without scheduling at least 75 hours of overtime. The functional team leader faces a dilemma: on the one hand, overtime pay was not budgeted for the project; on the other hand, there is substantial pressure to complete this contract on time, because future contracts with this client could be contingent on timely delivery. The team leader must make a decision.

In this case, as in all decision-making situations, it is important to make the best decision. How do team leaders know they have made the right decision? In most cases there is no *one* right choice. If there were, decision making would be easy. Typically there are several alternatives, each with its advantages and disadvantages.

In the case of XYZ Company, the team leader has two alternatives: authorize 75 hours of overtime or miss a deadline and risk losing future contracts. If the team leader authorizes the overtime, his company's profit margin for this project will suffer, but its relationship with the client will be preserved. If the team leader does not authorize the overtime, his company's planned profit will be protected, but its relationship with the client may be damaged.

Because it is not always clear at the outset what the best decision is, team leaders should be prepared to have their decisions criticized after the fact. It may seem unfair that managers, or employees, in the calm aftermath, criticize decisions made during the heat of the crisis. Having one's decisions evaluated is part of the accountability process, however, and can be an effective way to improve a team leader's decision-making skills.

Evaluating Decisions

There are two ways to evaluate decisions. The first is to examine the results. The result of a decision should advance a team toward accomplishing its goals, and to the extent that it does, the decision is usually considered good. Regardless of results, however, it is also wise to evaluate the process used in making a decision. A positive result can cause one to overlook the fact that you used a faulty process. And in the long run, faulty processes lead to more negative results than positive ones.

For example, consider a situation in which a team leader must choose among five alternatives. Rather than collecting as much information as possible, weighing the advantages and disadvantages of each alternative, and soliciting informed input, the team leader simply chooses randomly. There is one chance in five that she will choose the best alternative. Such odds will occasionally produce a positive result, but they typically will not.

It is important to examine the decision-making process as well as the decision's result—not just when the result is negative, but also when it is positive.

PROBLEMS AND DECISION MAKING

Everyone has problems—at home, at work, in relationships, and in every other facet of life. What constitutes a problem? Ask any number of people to describe their biggest problem and you will probably get as many responses. Here is a list of possible responses:

- "I don't make enough money."
- "I am overweight."
- "I need more education."
- "I don't like my job."

What these responses have in common is that they point to a difference between what is desired and what actually exists. A "problem" can be defined as follows:

> *A **problem** is the condition in which there is a discrepancy or a potential discrepancy between what is desired and what actually exists.*

Obviously the greater the disparity, the bigger the problem—with one exception. A key factor in determining the magnitude of a problem is the ability of the person with the problem to solve it. Even a pronounced disparity between what is desired and what exists does not represent a major problem if the person involved is able to eliminate the disparity. Correspondingly, even seemingly small disparities can represent big problems for people who do not have the ability or resources to eliminate them.

To illustrate this point, consider the following example: A new machine has been installed that can do five times as much work per hour as the one it replaced. But there is a problem: nobody can remember the correct start-up sequence. Using the wrong sequence could damage the machine. For most operators, this would be a small problem. They would simply consult the operator's manual, read the proper sequence, and follow the specifications. For the operator who cannot read, however, this would be a major problem. The difference is in the ability of the operator to solve the problem.

TEAMWORK TIP

"When I've heard all I need to make a decision, I don't take a vote. I make a decision."

—Ronald Reagan

Characteristics of Problems

Problems can be classified according to the degree of the following characteristics: *structure, organizational level,* and *urgency* (see Figure 7.1). The structure of a problem can vary from highly structured to no structure. A highly structured problem exists when the decision maker understands both the problem and how to solve it. An unstructured problem exists when the decision maker is unsure about alternatives and solutions. Highly structured and unstructured problems represent extremes. Problems with varying degrees of structure fall at different points along the continuum connecting these extremes.

Highly structured problems are so predictable that decisions can be automatic. For example, when a copy machine runs out of paper, the solution is obvious: add paper.

Unstructured problems are not predictable nor are responses to them automatic. For example, a team leader in a small business who temporarily loses her three best technicians when their Army Reserve unit is called to participate in a classified operation and does not know where they are going or when they will return. Consequently, she does not know whether to defer action and hope they will return soon, put her remaining staff on overtime, advertise for temporary employees, or request an extension on the contract her team is trying to complete. As is always the case with unstructured problems, this team leader needs to consider all alternatives carefully and seek informed input before making a decision.

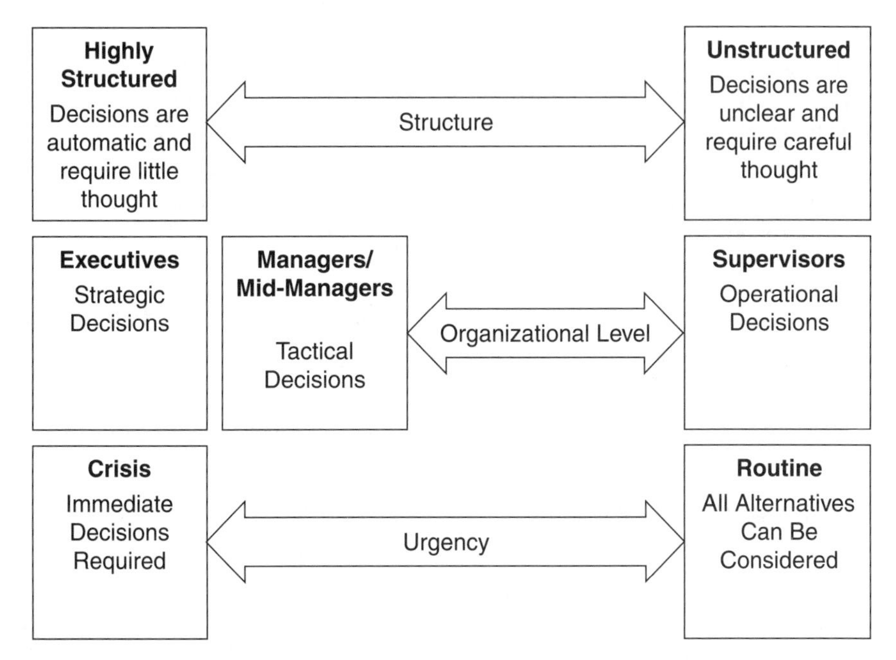

FIGURE 7.1 Decisions can be classified according to the degree of structure, organizational level, and urgency.

Problems also vary according to organizational level. Executive-level decision makers deal with strategic problems. Team leaders deal with operational-level problems—those that affect the day-to-day work of the organization. Team leaders are able to deal with problems within their span of authority without consulting higher management. Problems at the strategic level are outside the scope of their authority.

Problems can also be classified according to their degree of urgency. From this perspective, problems range from routine to those representing a crisis. One of the purposes of the management functions of planning and organization is to minimize the number of crisis problems that arise.

Crisis problems require immediate attention, forcing decision makers to react. Routine problems, on the other hand, give decision makers time to study the situation, consider alternatives, brainstorm ideas, and make well-reasoned decisions. In decision making it is always better to act than to react. For this reason, team leaders should apply the three-step approach to minimizing crisis problems: (1) plan, (2) organize, and (3) learn.

Careful planning and thorough organization will minimize the number of crisis problems team leaders must deal with. Even the best planning and organization will not completely eliminate crisis problems, however. This is why the third step—learn—is so important. Team leaders should learn from every crisis. They should ask questions such as, "Could this problem have been prevented? If so, how? Did I or my organization contribute to causing the crisis? Can changes be made to prevent a similar crisis in the future?" Team-members can contribute input to these questions as well.

Careful after-the-fact analysis of a crisis yields two important benefits. First, it can lead to improved planning and organization processes, thereby minimizing even further the number of future crises. Second, it can prepare team leaders to better handle a similar future crisis should one occur.

THE DECISION-MAKING PROCESS FOR TEAMS

Decision making involves a process. This is an important point to stress in a team setting.

> *The **decision-making process** is a logically sequenced series of activities through which decisions are made.*

There are numerous decision-making models. The one described in Figure 7.2 approaches every decision as a problem to be solved and divides the decision-making process into three distinct steps:

1. Identify/anticipate the problem.
2. Consider the alternatives.
3. Choose the best alternative, implement, monitor, and adjust.

Identify/Anticipate the Problem

If team leaders can anticipate problems, they may be able to prevent them. Anticipating problems is like driving defensively; you never assume anything. You look, listen, ask, and sense. For example, if you hear through the grapevine that a key employee was injured in the company's weekend softball game, you can anticipate the possible related problems. He may be absent, or if he can still work, his pace may be slowed. Knowing this, you can take the appropriate steps to fill in for this employee. The better team leaders know their team members, systems, products, and processes, the better they will be able to anticipate problems.

Even the most perceptive team leader will not always be able to anticipate problems. For example, if a team leader notices a "Who cares?" attitude among his team members, he might identify the problem as lagging employee morale and begin trying to make improvements. He would do well to confirm his hunch as to what is behind the negative attitudes, however. It could be that employees are upset about an unpopular management policy.

The processes at the heart of a problem might include those for which a team leader is responsible, such as scheduling and work processes, as well

Decision-Making/Problem-Solving Process

Identify/Anticipate the Problem

- Don't wait for problems to occur; look for them.
- Identify the root cause of the problem. Don't just treat the symptoms.

Consider the Alternatives

- Cost/benefit analysis
- Time considerations
- Permanent solution or temporary fix?
- Ethical considerations
- Moral considerations
- Think creatively

Choose the Best Alternative, Implement, Monitor, and Adjust

- Implement the best solution.
- Evaluate the results.
- Adjust as necessary.

FIGURE 7.2 Tasks involved in making decisions.

as a variety of others he does not control. These could include purchasing, inventory, delivery of material from suppliers, in-house delivery, quality control, and work methods. For example, if a team is having problems meeting production deadlines, the team leader might suspect sandbagging on the part of employees and take steps to solve that problem. The real problem could be the need to redesign the production process, however—a responsibility outside his control.

Resources that might be at the heart of a team leader's problem include time, money, supplies, material, personnel, and equipment. Is the problem caused by a lack of time? Insufficient funding? Poorly trained personnel? Outdated equipment? New, highly technical equipment that employees have not yet learned to operate proficiently? Poor-quality materials? All are possible causes of the problems with which team-leaders commonly deal.

The purpose of the lists in Figure 7.3 is to help team leaders look beneath the surface when attempting to identify the cause of a problem. This exercise can save time and energy that might otherwise be wasted dealing with symptoms rather than actual causes.

People and Processes	
■ Higher management (policies)	■ Inventory
■ Resource personnel	■ In-delivery
■ Suppliers	■ Out-shipping
■ Clients	■ In-work processes
■ Employees	■ Quality control
■ Training	■ Methods
■ Purchasing	■ Scheduling
Resources	
■ Time	■ Equipment/technology
■ Money	■ Material
■ Supplies	

FIGURE 7.3 Potential causes of problems.

Consider Alternatives

This is a two-step process. The first step is to list all the available alternatives to the current situation. The second is to evaluate each alternative. The number of alternatives identified in the first step is limited by several factors, including practical considerations, the team leader's range of authority, and the cause of the problem. Once the list has been developed, each entry on it is evaluated against the criterion of the desired outcome. If the problem is that a client's order is not going to be completed on time, will the alternative being considered solve the problem?

Cost is another important criterion used in evaluating alternatives. At what cost will each alternative solve the problem? Cost can be expressed in terms of finances, employee morale, the team's image, or a client's good will.

In addition to objective criteria, team leaders need to apply their experience, judgement, and intuition when considering alternatives.

Choose the Best Alternative, Monitor, and Adjust

Once all the alternatives have been considered, one must be selected and implemented. Team leaders should then monitor progress of the implementation and adjust it appropriately. Is the alternative having the desired effect? If not, what adjustments should be made?

Selecting the best alternative is an inexact process requiring logic, reason, intuition, guesswork, and luck. Occasionally, the alternative chosen for implementation does not produce the desired results. When this happens and adjustments are not effective, team leaders should cut their losses and move on to another alternative. As the adage goes, "If the horse you are riding dies, get off and find another."

Team leaders need to avoid falling into the "ownership trap." This happens when the leader invests so much of herself in a given alternative that she refuses to change even when it becomes clear that the idea is not working. This is most likely to happen when the selected alternative (1) runs counter to the advice the leader has received, (2) is unconventional, or (3) is unpopular. Remember, a team leader's job is to solve the problem. Feeling too much ownership in a given alternative can impede one's ability to do that.

DECISION-MAKING MODELS FOR TEAMS

Most of the many decision-making models available to team leaders fall into one of two categories: objective or subjective. In practice, a model may have characteristics of both.

Objective Approach to Decision Making

The objective approach is logical and orderly. It proceeds in a step-by-step manner and assumes that team leaders have the time to systematically pursue all the steps in the decision-making process. It also assumes that complete and accurate information is available and that team leaders are free to select what they feel is the best alternative.

Because these conditions rarely exist, a completely objective approach to decision-making is infrequently used. Team leaders seldom, if ever, have the luxury of time and complete information. This does not mean they should rule out objectivity in decision making, however—quite the contrary. Team leaders should be as objective as possible, recognizing that day-to-day realities of the workplace may limit the amount of time and information available. The degree of objectivity decreases correspondingly with the extent of these limitations.

Subjective Approach to Decision Making

Whereas the objective approach to decision making is based on logic and complete, accurate information, the subjective approach is based on intuition, experience, and incomplete information. This approach assumes decision makers are under pressure, short on time, and operating with limited information. The goal of subjective decision making is to arrive at the best possible decision under the circumstances.

With this approach, there is always the danger that team leaders will make quick, knee-jerk decisions based on little information or input from other sources. The subjective approach does not give team leaders license to make careless decisions. If time is short, use the little time available to list and evaluate alternatives. If information is incomplete, use as much information as you have. Then call on your experience and intuition to fill in the rest of the picture. Never fall back on subjective decision making to make up for laziness in collecting accurate information or failure to budget time wisely.

EMPOWERING EMPLOYEES IN DECISION MAKING

Decision making can be improved by involving the employees who will have to carry out the decision or who are affected by it. Employees are more likely to feel ownership in a decision they had a part in making, and they are more likely to support a decision for which they feel ownership. Collecting input from employees before making decisions that affect them and that they will have to carry out is called "empowerment." There are both advantages and disadvantages to involving and empowering employees in decision making.

Advantages of Employee Empowerment

Involving and empowering employees in decision making can result in a more accurate picture of the problem and a more comprehensive list of alternatives. It can help team leaders better evaluate alternatives and select the best one to implement. Perhaps the most important advantages are gained after the decision is made. Employees who participate in the decision-making process are more likely to understand and accept the decision, and they have a personal stake in ensuring the success of the selected alternative.

Disadvantages of Employee Empowerment

Involving and empowering employees in decision making can take time, and team leaders do not always have the requisite time. In addition, empowerment in decision making takes employees away from their jobs and can result in conflict among team members. Another significant disadvantage is that it can lead to democratic compromises that do not necessarily represent the best decision. Disharmony can result if the team leader rejects the advice of the group.

Several techniques can help increase the effectiveness of group involvement. Prominent among these are brainstorming, the nominal-group technique, and quality circles.

Brainstorming

With *brainstorming*, the team leader serves as a catalyst in drawing out group members to share any idea that comes to mind. All ideas are considered valid. Team members are not allowed to make judgmental comments or to evaluate any suggestion that is made. Typically, one member of the team is asked to serve as a recorder. All ideas are recorded, preferably on a marker board, flip chart, or another medium that allows team members to review them continuously.

Once all ideas have been recorded, participants are asked to go through the list one item at a time and weigh the relative merits of each. This process is repeated until the group narrows the choices to a specified number. A team leader may ask the group to narrow the number of alternatives down to three, for instance, reserving the selection of the best of the three for himself.

Nominal-Group Technique

The *nominal-group technique (NGT)* is a sophisticated form of brainstorming. It has the following five steps:

1. The problem is stated.
2. Team members silently record ideas.
3. The ideas of each member are reported publicly.
4. The ideas are clarified.
5. The ideas are silently voted on.

In the first step, the team leader states the problem and makes clarifications, if necessary, to make sure all team members understand it. In the second step, each team member silently records his ideas. At this point, there is no discussion among team members. This strategy promotes free and open thinking unimpeded by judgmental comments or peer pressure.

In the third step, each team member shares one idea with the group. The ideas are recorded on a marker board or flip chart. The process is repeated until all ideas have been recorded. Each idea is numbered. During this step as well, there is no discussion among team members. Taking the ideas in rounds of one at a time from each team member ensures a mix of recorded ideas, making it more difficult for participants to remember which ideas belong to which member. In the fourth step any ideas that are not understood are clarified.

In the fifth step, the ideas are voted on silently. There are a number of ways to do this. One simple technique is to ask all team members to record the numbers of their five favorite ideas on five separate 3″ × 5″ cards and then to prioritize the five cards by assigning them numbers ranging from 1 to 5, with 5 representing the best idea.

Quality Circles

A *quality circle* is a group of employees convened to solve problems relating to their jobs. The underlying principle of the quality circle is that the people who do the work know the most about the work and should therefore be involved in solving work-related problems. A key difference between a brainstorming group and a quality circle is that members of the latter are volunteers who convene themselves without being directed to do so by the team leader. In addition, they don't wait for a problem to occur before they meet. Rather, they confer regularly to discuss their work, anticipate problems, and identify ways to improve productivity. A quality circle has a person—not necessarily the team leader—who acts as the facilitator. It may be a different person for each meeting.

Potential Problems with Group Decision Making

Team leaders interested in improving decision making through group techniques should be familiar with the concepts of *groupthink* and *groupshift.* These two phenomena can undermine the effectiveness of the group techniques set forth in this section.

Groupthink occurs when people in a group focus more on reaching a decision than on making the right decision. A number of factors can contribute to groupthink, including the following: overly prescriptive group leadership, peer pressure for conformity, group isolation, and unskilled application of group decision-making techniques. The following strategies can be used to overcome groupthink:

- Encourage criticism.
- Encourage the development of several alternatives. Do not allow the group to rush to a hasty decision.
- Assign one or more members to play the role of devil's advocate.
- Include people who are not familiar with the issue.
- Hold a "last-chance meeting." Once a decision is reached and group members have had time to think it over, call a last-chance meeting, in case group members are having second thoughts.

Groupshift occurs when team members exaggerate their initial position, hoping that the eventual decision will be what they really wanted in the first place. If team members decide ahead of time to take an overly risky or overly conservative initial view, it can be difficult to overcome the resulting groupshift. Leaders can help minimize the effects of groupshift by discouraging reinforcement of initial points of view and by assigning team members to serve as devil's advocates.

THE ROLE OF INFORMATION IN DECISION MAKING

Information is a critical element in decision making. Although accurate, up-to-date, comprehensive information does not guarantee a good decision, the lack of it can guarantee a bad one. The old saying "knowledge is power" applies in decision making, particularly in a competitive situation.

"Information" can be defined as data, converted into a usable format, that are relevant to the decision-making process. Recall that communication requires a sender, a medium, and a receiver. In the communication process, information is sent by the sender, transmitted by the medium, and received by the receiver. For the purpose of this book, team leaders are receivers of information who base decisions at least in part on that information.

Advances in technology have given team leaders instant access to information. Of course, the quality of that information depends on the accuracy of the data entered into these technological systems and on continuous updating of that data. The information age gave rise to the saying "garbage in—garbage out"; that is, the information provided by a computer-based information system can be no better than the data put into it.

Data versus Information

Data for one person may be information for another. The difference lies in the needs of the individual. A team leader's needs are dictated by the types of decisions he is required to make. In deciding on the type of information they need, team leaders should ask themselves these questions:

- What are my responsibilities?
- What are my organizational goals?
- What types of decisions do I have to make relative to these responsibilities and goals?

The Value of Information

Information is a useful commodity. As such, it has value, as determined by the needs of the people who will use it and the extent to which it will help them meet their needs. Information also has a cost. Because it must be collected, stored, processed, continually updated, and presented in a usable format when needed, information can be expensive. Therefore, team leaders must weigh the value of information against its cost when deciding on what information they need to make a decision. It makes no sense to spend $100 on the information needed to make a $10 decision.

Amount of Information

There was a saying that went like this: "A manager can't have too much information." This is no longer true. With advances in information technology, not only can managers have too much information, they frequently do. This phenomenon has come to be known as *information overload,* the condition that exists when people receive more information than they can process in a timely manner. "In a timely manner" means in time to be useful in decision making. Information overload can cause the following problems:

- Confusion
- Frustration
- Excessive attention given to unimportant matters
- Insufficient attention given to important matters
- Unproductive delays in decision making

To avoid information overload, team leaders can apply the following strategies:

- Change daily reports to weekly reports wherever possible.
- Change weekly reports to monthly reports wherever possible.
- Examine all reports provided on a regular basis, and eliminate those that are unnecessary.
- Apply the "reporting-by-exception" technique.
- Format reports for efficiency.
- Use online, on-demand information retrieval.

Examine all reports you receive on a regular basis. Are they really necessary? Do you receive daily or weekly reports that would meet your needs just as well if provided on a monthly basis? Do you receive regular reports that would meet your needs better as exception reports? In other words, would you rather receive daily reports that say "Everything is all right" or occasional reports only when there is a problem? Choosing the latter type of report can cut down significantly on the amount of information team leaders must deal with.

The formatting for efficiency strategy involves working with personnel who provide information, such as management information systems (MIS) employees, to ensure that reports are formatted for your convenience, not theirs. Team leaders should not wade through reams of computer printouts to locate the information they need, nor should they become bleary-eyed reading rows and columns of tiny figures. MIS personnel can recommend an efficient report format to meet your needs. Also, information can be presented graphically whenever possible.

Finally, make use of online, on-demand information retrieval. In the modern work setting, most reports are computer generated. Rather than relying on periodic hard-copy reports, learn to retrieve information online.

THE ROLE OF CREATIVITY IN DECISION MAKING

The increasing pressures of a competitive marketplace are making it more and more important for organizations to be flexible, innovative, and creative in their decision making. To survive in an unsure, rapidly changing marketplace, organizations must be able to adjust rapidly and change directions quickly. This requires creativity at all levels of the organization, including the team level.

Creativity Defined

As we saw with "leadership," "creativity" has many definitions, and there are varying viewpoints concerning whether creative people are born or made. For the purpose of global organizations, a useful definition is:

> ***Creativity*** *is the process of developing new, different, imaginative, and innovative perspectives when dealing with problems that affect any aspect of competitiveness.*

Developing such perspectives requires that decision makers have knowledge and experience regarding the issue in question.

Creative Process

The creative process proceeds in three stages: investigation, incubation, and clarification.

- *Investigation* involves learning, gaining experience, and collecting/storing information about a given area. Creative decision making requires that the people involved be thoroughly prepared through investigation and research.
- *Incubation* means giving ideas time to develop, change, grow, and solidify. Ideas incubate while decision makers drive, relax, sleep, and ponder. Incubation requires that decision makers get away from the issue in question and give the mind time to sort things out. Incubation is often accomplished by the subconscious mind.
- *Clarification* follows incubation. It is the point at which a potential solution becomes clear. This point is sometimes the moment of inspiration. However, inspiration rarely occurs without having been preceded by clarification to determine if the solution will actually work. At this point, traditional processes such as feasibility studies and cost-benefit analyses are used to evaluate the decision.

Factors That Inhibit Creativity

The following factors tend to inhibit the creative process:

- *Looking for the one right answer.* There is seldom just one right solution to a problem.
- *Focusing too intently on being logical.* Creative solutions sometimes defy perceived logic and conventional wisdom.
- *Adhering too closely to the rules.* Sometimes the best solutions come from stepping outside the box and looking beyond the limits established by prevailing rules.
- *Focusing too intently on practicality.* Impractical ideas can sometimes trigger practical solutions.
- *Avoiding ambiguity.* Ambiguity is a normal part of the creative process. This is why the incubation step is so important.
- *Avoiding risk.* When teams can't find a solution to a problem, it often means decision makers are not willing to take a chance on a creative decision.
- *Forgetting how to play.* Adults sometimes become so serious they forget how to play. Playful activity can stimulate creative ideas.
- *Fear of rejection or looking foolish.* This fear can cause people to hold back what might be creative solutions.
- *Saying "I'm not creative."* People who decide they are not creative won't be. Any person can think creatively and can learn to be even more creative.

TEAMWORK TIP

"If the only tool you have is a hammer, you tend to see every problem as a nail."

—Abraham Maslow, American psychologist

Summary

1. Decision making is the process of choosing one alternative from among two or more alternatives. Ideally the alternative chosen is the best or at least the optimal alternative available.

2. A problem is the condition that exists when there is a discrepancy or a potential discrepancy between what is desired and what actually exists. Problems have the following characteristics: structure, organizational level, and urgency.

3. The decision-making process for teams is a logically sequenced series of activities through which decisions are made. It has three distinct steps:

identify/anticipate the problem; consider the alternatives; and choose the best alternative, implement, monitor, and adjust. The two broad approaches to decision making are the objective approach and the subjective approach.

4. Collecting input from employees before making decisions that affect them is called empowerment. The principal disadvantage of empowerment is that it takes time. The principal advantage is buy-in. Empowered employees are more likely to accept and support (buy into) decisions because they had a voice in making them.

5. Brainstorming, the nominal-group technique, and quality circles are vehicles for soliciting employee input when making decisions. All three vehicles promote empowerment.

6. Information is data, converted into a useable format, that are relevant to the decision-making process. Data that are relevant to decision making have an impact on the decision.

7. Creativity is the process of developing new, different, imaginative, and innovative perspectives when dealing with problems that affect any aspect of competitiveness. The creative process proceeds in three stages: investigation, incubation, and clarification.

Key Terms and Concepts

Brainstorming
Clarification
Creativity
Crisis problems
Decision making
Decision-making process
Empowerment
Exception report
Groupshift
Groupthink
Incubation
Information
Information overload
Investigation
Nominal-group technique
Objective approach
Organizational level
Problem
Quality circle
Review
Routine problems
Structure
Subjective approach
Urgency

Review Questions

1. Define the term "decision making."
2. What is the concept of *right choice* as it relates to decision making?
3. What are two ways to evaluate decisions? Explain them briefly.

4. What is a problem?
5. What are the three characteristics of problems?
6. What is a crisis problem?
7. How can team leaders minimize the number of crisis problems they have to deal with?
8. Define the term "decision-making process."
9. What are the three steps in the decision-making process? Explain each briefly.
10. Compare and contrast the objective and subjective approaches to decision making.
11. What are the advantages of involving employees in decision making?
12. What are the disadvantages of involving employees in decision making?
13. Briefly explain the following strategies for involving employees in decision making: brainstorming, NGT, quality circles.
14. How can the concepts of groupthink and groupshift affect decisions made in groups?
15. Define the term "information."
16. What is the difference between data and information? What determines the difference?
17. What is information overload? How can team leaders avoid it?
18. List and explain the three stages of the creative process.

EFFECTIVE TEAMWORK SIMULATION CASES

The following cases deal with specific issues relating to the implementation of effective teamwork. Each case represents a meeting of Marcee McPhee and Pete Fared, engineers and team leaders at Mac-Tech, Inc. a technology firm with 526 employees. McPhee is the leader of Team A, and Fared leads Team B. McPhee and Fared are not just colleagues; they are friends, and their friendship goes all the way back to college. They both attended the same engineering school and graduated in the same class. Once a week they meet for lunch and discuss problems, progress, issues, and concerns. These cases chronicle their luncheon conversations and invite the reader to discuss the issues Fared and McPhee deal with.

CASE 7.1 Problem Solving Is a Process

"Do you have problems when you make a decision, Marcee?" asked Pete Fared. Lately, several of his decisions had turned out badly. "Why do you ask?" queried Marcee McPhee. "Well, I make decisions all the time in my team, but the results are erratic. Usually when I make decisions everything turns out just the way it should, but not always. Sometimes I make a decision and it doesn't work at all. Things actually become worse."

"All decision makers have that problem. No matter how hard you try, some decisions just don't work out," counseled McPhee. "I suppose," said Fared noncommittally. "But it just seems to me there must be a more systematic way that decisions can be made." "There is. Haven't you ever heard of the decision-making process?"

"I guess so," mumbled Fared unconvincingly. "But I never paid much attention to it. I always figured that team leaders make the decisions because we have the most knowledge and experience." "We do," acknowledged McPhee. "But that doesn't mean we just wing it. Making a decision should follow a systematic process. That way, if things don't work out, at least we can analyze the process to determine what went wrong."

"You keep saying that decision making involves a process," said Fared. "What is the process, and how does it work?"

Discussion Questions

1. Put yourself in Marcee McPhee's place in this scenario. How would you answer Pete Fared's last question?
2. Give a step-by-step example of how the decision-making process works.

CASE 7.2 Empowerment: Involving Employees in Decision Making

"So, Pete, has your decision making improved since you started using a systematic process?" asked Marcee McPhee. Pete Fared had stopped making seat-of-the-pants decisions and had started using a step-by-step process. "There's been some improvement," admitted Fared. "But there are still problems. The decision-making process isn't perfect."

"No it isn't," acknowledged McPhee. "But then, no process is. Tell me about the problems you are having." "Well, it seems like the quality of my decisions is determined at the front end of the process," began Fared. "Are you talking about the information collection?" asked McPhee. "Precisely," acknowledged Fared. "It's the old garbage in—garbage out. I don't always have time to run around collecting information before making a decision. At least half the time I wind up making a decision based on incomplete information," huffed Fared, clearly exasperated.

"That's always a problem," acknowledged McPhee. "Decision makers seldom have complete information, but we can get sufficient information if we go about it right." "How do you go about getting sufficient information before making decisions in your team?" asked Fared.

"That's easy. I involve my team members in the decision-making process. You'd be surprised at how much good information they can give you. It's called 'empowerment,' " said McPhee. "Tell me about empowerment and how I can use it to improve decision making."

Discussion Questions

1. Put yourself in Marcee McPhee's place in this scenario. How would you respond to Pete Fared's request for an explanation of empowerment and how to use it to improve decision making?
2. Give an example from your experience of how empowerment improved a specific decision.

CASE 7.3 Creativity in Decision Making

"Pete, has decision making improved since you started empowering team members?" asked Marcee McPhee. Pete Fared didn't answer his friend at first. He appeared lost in thought. Finally he said, "I was skeptical as first, but I have to admit that you were right. I can think of several times in the past two weeks when everyone's input really improved a decision I had to make."

"I'm glad to hear it," acknowledged McPhee. "That means you are ready for the next step." Fared cringed, holding up his hands as if to ward off bad news. "The next step? Don't tell me there's more!" "Of course there is," said McPhee with a spider-to-the-fly grin. "I should have known," admitted Fared with a sigh.

"You're going to love the next step," said McPhee assuringly. "It's all about creativity." "Creativity!" blurted Fared. "Come on, I'm an engineer, not an artist." "Now there you go with the negativity. That's just the kind of attitude that will kill creativity in your team," scolded McPhee. "Marcee, you and I are engineers. We are trained to examine a problem and find a solution. We use the scientific method, not creativity," declared Fared.

"Science and creativity are not mutually exclusive concepts," said McPhee with a touch of exasperation. "People who can't even paint by numbers, much less create works of art, can still be creative. It's just a matter of thinking outside the box, using your imagination, and being innovative when solving problems. But with your attitude, you'll never be creative," declared McPhee.

"All right, Marcee. I get the point," said Fared with a look of surrender. "What can I do to make sure I don't inhibit creativity in my team?"

Discussion Questions

1. Do you agree with Marcee McPhee that anybody can be creative? Explain your reasoning.
2. Put yourself in McPhee's place in this scenario. How would you answer Fared's last question?

8

Establish Positive Team Behaviors

The most important measure of how good a game I played was how much better I'd made my teammates play.

—Bill Russell

OBJECTIVES

Explain the following concepts:

- selflessness in teamwork.
- honesty and integrity in teamwork.
- dependability in teamwork.
- enthusiasm in teamwork.
- responsibility in teamwork.
- cooperativeness in teamwork.
- taking initiative in teamwork.
- patience in teamwork.
- resourcefulness in teamwork.

- punctuality in teamwork.
- tolerance and sensitivity in teamwork.
- perseverance in teamwork.

Leadership and teamwork expert John Maxwell says, "There is no I in teamwork."[1] This is a catchy and succinct way of summing up how essential selflessness is to teamwork. When it comes to effective teamwork, selflessness is an absolute prerequisite. "Selflessness" in teamwork is *a willingness to put the goals of the team ahead of your own personal goals.* This concept is easy to define but difficult to implement in teams or to act out in our personal behavior on a day-to-day basis.

SELFLESSNESS IN TEAMWORK

Most people admire selfless behavior when they see others display it. Admiring what others do is not the same as doing it yourself, however. To have any value in a team setting, selflessness must be more than an admirable concept. It must be proactive behavior displayed consistently. Some behaviors that convert the concept of selflessness into action are as follows:

- Give to other team members.
- Refuse to gossip.
- Avoid territorial behavior.
- Be loyal to teammates.

Give to Other Team Members

Few things build unity in a team faster or better than generosity. When one team member is generous to another, two things happen. First, a special bond begins to develop between the two team members. That bond will grow over time as it is reinforced by additional acts of mutual generosity. Second, a positive example is set for other team members. When team members observe their colleagues being generous to each other, peer pressure begins to work on behalf of generosity. Peer pressure is a powerful force in teams. With peer pressure positively applied, generosity begets more generosity.

One of the best ways to act generously toward fellow team members is to help them do their jobs better, offering your time, experience, and expertise—all valuable commodities. Some of the many other ways to show generosity to a team member include taking his place occasionally when he is needed to work overtime, finishing up her work now and then so she can attend a child's ball game or recital, and offering to work for him when he has a sick child at home.

Refuse to Gossip

Few things can have a more deleterious effect in a teamwork setting than gossip. Gossip has a doubly negative effect in that it hurts both the targeted individual and the team as a whole. Gossip hurts the targeted individual because she is maligned without having an opportunity to tell her side of the story or defend herself. Gossip hurts the team because it breaks down trust—the glue that holds teams together. Even if the gossip is about someone outside the team, the effect is still negative because team members who hear it will think, "If he gossips about her, he will gossip about me."

Avoid Territorial Behavior

Territorial behavior comes naturally to human beings; it does not have to be learned. Observe even the youngest children and you will see territorial behavior: "That's my toy! You can't play with it. It's mine, mine, mine!" Sharing, on the other hand, is not natural human behavior; it is learned. Ask any parent. The same types of territorial behavior observed in children are acted out every day by adults in the workplace. This behavior manifests itself in a variety of ways:[2]

- *Information manipulation.* People who play territorial games with information subscribe to the philosophy that information is power. To feel powerful, they withhold information, bias (spin) information to suit their individual agendas, cover up information, and give out false information.
- *Intimidation.* One of the most common manifestations of territoriality is intimidation—a tactic used to frighten others away from certain turf. Intimidation can take many forms, from subtle threats to blatant aggression (physical or verbal).
- *Alliances.* Forming alliances with powerful individuals in an organization is a commonly practiced territorial game. The idea is to communicate, without actually having to speak the words, "You had better keep off my turf, or I'll get my powerful friend to cause trouble."
- *Invisible walls.* The goal of putting up invisible walls is to ensure that a decision, although already made, cannot be implemented. There are hundreds of means of creating hidden barriers, including stalling, losing paperwork, and forgetting to place an order.
- *Strategic noncompliance.* Agreeing to a decision up front, but having no intention of carrying it out is called "strategic noncompliance." This tactic is often used to buy time to find a way to reverse the decision.
- *Discrediting.* Discrediting an individual to cast doubt on his or her recommendation is a common turf-protection tactic. Such an approach is called an "*ad hominem* argument," meaning if you cannot discredit the recommendation, discredit the person making it.

- *Shunning.* Excluding an individual, who threatens your turf is also a common territorial-protection tactic. Shunning creates peer pressure against the excluded individual.
- *Camouflage.* This tactic can also be called "throwing up a smoke screen" or "creating a fog." It involves confusing the issue by raising other, distracting controversies—especially those intended to produce anxiety.
- *Filibustering.* "Filibustering" means talking a recommended action to death. It involves talking at length about concerns—usually inconsequential ones—until either the other side gives in just to stop any further discussion or the time to make the decision runs out.

Overcoming territorial behavior requires a two-step approach: (1) recognizing one or more of the manifestations described earlier and admitting that they exist, and (2) creating an environment in which survival is equated with cooperation rather than territoriality. Simmons recommends the following strategies for creating a cooperative environment:[3]

- *Avoid jumping to conclusions about the motives of people.* Talk to team members about territoriality versus cooperation. Ask to hear their views, and listen to what they say.
- *Attribute territorial behavior to instinct rather than to people.* Blaming people for following their natural instincts is like blaming them for eating. The better approach is to show them that their survival instinct is tied to cooperation, not territoriality. This is done by rewarding cooperation and applying negative reinforcement to territorial behavior.
- *Ensure that no team member feels attacked.* Remember that the survival instinct is the motivation behind territorial behavior. Attacking team members, or even leading them to feel attacked, triggers their survival instinct. To change territorial behavior it is necessary to put employees at ease.
- *Avoid generalizations.* When a team member exhibits territorial behavior, deal with the specific situation as opposed to generalizing it. It is a mistake to witness territorial behavior on the part of one team member and to respond to it on a group level. Deal with the individual who exhibits the behavior, and focus on specifics.
- *Understand irrational fears.* The survival instinct is a powerful motivator that can cause team members to cling to irrational fears. Team leaders should consider this point when dealing with team members who find it difficult to let go of survival behaviors. Be firm but patient, and never deal with a team member's fears in a denigrating or condescending manner.
- *Respect each individual's perspective.* In a way, an individual's perspective or opinion is part of his psychological territory. Failure to respect the team member's perspectives is the same as threatening their territory. When challenging territorial behavior, let employees explain their perspectives, and show respect for them even if you do not agree.

- *Consider the employee's point of view.* In addition to demonstrating respect for the team member's perspectives, team leaders should try to "step into their shoes." How would you feel if you were the team member? Sensitivity to and patience with the employee's point of view are critical when trying to overcome territorial behavior.

Be Loyal to Teammates

Loyal teammates can be counted on to protect the best interests of the team. Self-interest, rather than loyalty, is too often the norm when people interact in groups. Consequently, when teammates display loyalty to each other, they are breaking the mold in a positive sense. Loyal teammates take care of each other, help each other, and improve each other.

The military provides one of the best sources of examples demonstrating the value of loyalty in teams. Historical literature is replete with examples of selfless behavior on the part of military personnel who risked their lives out of loyalty to their team members. One such example comes from the battle for the tiny but strategically crucial Pacific island of Iwo Jima during World War II.

Most people are familiar with the monument to the Marines who, during the battle for Iwo Jima, raised the American flag atop Mount Surabachi. The moving inscription on the base of this monument says, "Uncommon valor was a common virtue," which is to say that acts of loyalty and courage during the battle were too many to count.

There were six young men captured on film raising the American flag on Mount Surabachi in Joe Rosenthal's Pulitzer-Prize-winning photograph that was the basis for the monument. Look closely at the photograph and you will notice that the flag raiser wears no weapon of any kind. This is because he is not a Marine but a Navy medic serving with the Marines. This man, John Bradley, although not a Marine, was an important member of a Marine platoon. In the slaughterhouse that was Iwo Jima, Bradley risked his life over and over to administer first aid to the injured and to save the lives of wounded Marines (his teammates). By the time the Japanese soldiers defending Iwo Jima had been defeated, John Bradley had been wounded twice himself, but he did not stop caring for his teammates until he and they were evacuated from the island. Bradley's quiet but courageous loyalty inspired many Marines at Iwo Jima to persevere through the battle to a costly but critical victory.

HONESTY AND INTEGRITY IN TEAMWORK

Teamwork is built on trust, and trust is built on honesty and integrity. Put another way, honesty and integrity are the foundation of trust, and trust is the foundation of teamwork. Teammates depend on each other every day in ways that affect them personally, such as job performance, job security, wages, and promotions. It is difficult, under even the best of circumstances, for people to place their personal interests in the hands of others, even if only partially. Trust, honesty, and integrity make this possible.

TEAMWORK TIP

"I have found that being honest is the best technique I can use. Right up front, tell people what you're trying to accomplish and what you're willing to sacrifice to accomplish it."

—Lee Iacocca, American entrepreneur and business leader

DEPENDABILITY IN TEAMWORK

Dependable people consistently do what they are supposed to do when they are supposed to do it and in the way they are supposed to do it. Because team members rely on each other to accomplish their work, dependability is critical to effective teamwork. In brief, the performance of the team depends on the performance of its members.

Dependability means more than just doing one's best. It means doing what is necessary to get the job done within the bounds of ethical behavior. Dependable team members have the following characteristics:

- They follow through on promises and expectations.
- They are punctual.
- They take responsibility and are accountable.
- They complete what they start.
- They chip in to help others.

ENTHUSIASM IN TEAMWORK

The concepts of team spirit and *espirit de corps* are real. People who are enthusiastic about their work typically do it better. This is not to say that competence is irrelevant—quite the contrary. Enthusiasm without competence is like fire without light. Enthusiasm can multiply the positive effects of competence, however.

The good news is that enthusiasm is contagious. The bad news is that negativism is also contagious. Every team faces difficulties and barriers. Enthusiasm can help team members persevere when the road gets rocky. Negativism, on the other hand, promotes a defeatist attitude that says, "When the going gets tough, give up." The best defense against negativism is enthusiasm. Leadership expert Dennis Waitley has this to say about enthusiasm:

"Enthusiasm is contagious. It's difficult to remain neutral or indifferent in the presence of a positive thinker."[4]

It's easy to be enthusiastic when things are going your way or when the work you do is exciting and fun, but things don't always go your way and work isn't always exciting and fun—even for those who love their jobs. The challenge inherent in teamwork is to be enthusiastic about the work even when you don't feel like doing it, a task more easily said than done. The following strategies can help you be enthusiastic in the face of adversity, boredom, or other work-related difficulties:

- *Choose to be enthusiastic.* As odd as it may sound, we actually choose whether to face the world with enthusiasm. Those who find this difficult to believe confuse emotion with attitude. Sadness and anger are emotional responses to the exigencies of life. We do not choose the emotions we feel, although we can choose how we respond to them. Enthusiasm, on the other hand, is an *attitude,* not an emotion. Whereas an emotion is an involuntary response to the exigencies of life, an attitude is a chosen response. For example, if a team leader gives you a boring assignment, you might feel angry, but you can still choose to put your anger aside and approach the assignment with enthusiasm.
- *Think positively.* A positive attitude engenders enthusiasm even when you don't feel enthusiastic. Remember that your attitude is a choice, so you can choose to have a positive attitude. If you act in a positive way, your unconscious emotions sometimes catch up with your conscious actions, and you will begin to feel positive and enthusiastic.
- *Seek out enthusiastic people.* Certain behavior is contagious. Just try to keep a straight face when everyone else in the room is laughing! Enthusiasm is one of those contagious behaviors. For most people, it is difficult, if not impossible, to remain dull, impassive, lethargic, or bored around someone who is enthusiastic. Consequently, when trying to stay upbeat yourself, it is a good idea to spend time with people who are enthusiastic. The reverse is also true. Negativity is contagious. So when you are trying to maintain your enthusiasm, make a point of limiting the amount of time you spend with negative people.

RESPONSIBILITY IN TEAMWORK

Responsible team members make a point of knowing what needs to be done and either doing it themselves or making sure it gets done. In addition, responsible team members expect to be held accountable for getting the job done right and on time. The most effective teams are comprised of members who take responsibility for their actions, decisions, and performance, as well as for those of the team. Avoidance of responsibility is a common characteristic of members of ineffective teams. When things go wrong in such teams, members typically blame each other or look to find a convenient scapegoat.

Responsible and accountable employees hold a team together when difficulties arise. On the other hand, avoiding responsibility and shrinking from accountability causes a team to fall apart during difficult times.

The following example illustrates the contribution that responsible employees can make to a company. A package-delivery company in a large city had found its niche making intracity deliveries for technology companies. The company hired local college students who were familiar with the city's streets, avenues, and alleys to work part time picking up and delivering packages to the many technology companies located there. One morning, John, the deliverer who worked the 8:00 A.M. to 10:00 A.M. shift, was given a package with a delivery ticket marked *"Urgent—Rush!"* John put the package on top of his stack and dashed off to make his deliveries.

TEAMWORK TIP

"You can't escape the responsibility of tomorrow by evading it today."

—Abraham Lincoln, 16th president of the United States

Upon arriving at the address on his delivery ticket, John found nothing but an empty building that was clearly not a technology company and did not appear ever to have been one. With several other packages to deliver during his short, two-hour shift, John shrugged, tossed the urgent package back in the delivery van, and sped away on his rounds. By the time John had delivered all his other packages, his shift was almost over and he barely had time to make it to campus for his first class of the day. John put the "urgent" package on top of the stack awaiting the deliverer who worked the 10:00 A.M. to noon shift, and scribbled a quick note on the delivery ticket that said, "wrong address."

Mary noticed the urgent package when she picked up her delivery tickets, and she placed it on top of the other packages she had to deliver. Clearly the urgent package had priority. She would deliver it first. Mary noted that the package had a wrong address. The writing on the delivery ticket had been hastily scrawled, but she thought she could make it out. She looked up the address of the recipient company, or at least what she thought was the recipient company, in the business and industry directory that was always on the front seat of the delivery van. She wrote down the correct address and drove as fast as the law and her concern for safety would allow to the address in question. Mary found the lights out, the door locked, and a small "Out of business" sign taped inside a window.

Mary found it odd that an urgent package would be addressed to a company that had gone out of business, but she didn't dwell on the issue because there were other packages to deliver. When she returned the delivery van at the end of her shift, Don, the noon to 2:00 P.M. deliverer, was waiting for her. She handed Don the urgent package and explained its short but strange history.

Don knew immediately that something was wrong, but he did not know what it could be. Somebody needed this package, and they needed it right away. He had a full slate of deliveries to make by 2:00 P.M., but as he drove, the thought of the urgent package kept gnawing at him. Somebody needed that package, and Don decided he would find out who and make the delivery, no matter how long it took him.

It was 1:45 P.M. when Don completed all of his other deliveries. His shift would be over in just 15 minutes, and he had a 2:30 P.M. class at the university that day. No matter. Don decided he was going to deliver the urgent package even if it meant missing his class, which, as things turned out, it did.

Don looked at the name of the recipient company on the delivery ticket. It had been hastily scribbled in longhand and was difficult to read. It could say "Mera-Tech," "Mana-Tech," or "Meva-Tech." Don couldn't tell which was correct. He took the business and industry directory out of the delivery van and began looking under the M's. There was a Mana-Tech listed, but according to the address in the directory, this was the company Mary had found to be out of business. Next, Don tried Mera-Tech. There was a listing, but the address was nothing like the one on the delivery ticket. Then Don tried Meva-Tech. There was a listing, and the address looked similar to the one John had marked as being a wrong address, but with a transposed number. Don called the telephone number listed for Meva-Tech and asked the receptionist if the company was expecting an urgent package.

"Thank goodness. Where have you been?" asked Meva-Tech's receptionist, practically shouting. She told Don that the package contained important papers for a meeting scheduled to begin in just 10 minutes. "Without those papers, my company is toast," claimed the frantic receptionist. Don couldn't take the delivery van because in just a few minutes his replacement deliverer would show up for work and would need it. But Meva-Tech's office was just six blocks away. Don told the receptionist not to worry, he would be there in five minutes—and he was. Completely out of breath from his exertion, Don handed the package to a much-relieved and appreciative receptionist.

The receptionist, Margaret, asked Don to wait while she delivered the package to the company's conference room, which she did with three minutes to spare. Then she took Don aside and asked why his company had cut it so close on delivering the package. Don apologized on behalf of his employer and told Margaret the whole convoluted story. Knowing that the delivery company hired college students, Margaret asked Don what he was majoring in. He said, "electrical engineering," and dashed out the door, hoping to make it to the university in time to catch up with his professor and explain why he had missed class.

Later that day, Margaret told her boss the story of how Meva-Tech's package was delivered at the last minute by an enterprising college student. The next day, when Don showed up for work he found an envelope attached to his delivery tickets. The note inside was from the vice-president for engineering at Meva-Tech. It read, "When you graduate with your engineering degree, there is a job for you at Meva-Tech. I need people on my team who know how to take responsibility for getting the job done."

COOPERATIVENESS IN TEAMWORK

For teams to work effectively, their members must work cooperatively with each other. Think of track teams that run relay races. Much of their success depends on how they cooperate in passing the baton at each hand off. If team members fail to cooperate in this critical phase of the race, the team is likely to lose, no matter how fast each individual member runs his leg of the relay.

This need for cooperation among team members exists in every kind of team. Observe the cooperation needed among the center, holder, and place kicker to score a field goal in football. Observe the cooperation needed among the shortstop, second baseman, and first baseman to make a double play in baseball. In all of these examples, each team member must cooperate by doing his job in such way as to ensure that the next player can do his. This is as critical in work teams as it is in sports teams.

INITIATIVE IN TEAMWORK

Initiative means recognizing what needs to be done and doing it without waiting to be told. In teams where initiative is the norm, one never hears team members say, "That's not my job." Instead, one observes people who get their jobs done and then look for ways to help their teammates get their jobs done.

Team members with initiative approach their jobs as if they are team leaders. They consistently look beyond their specific duties to find ways to make the team as a whole perform better. When they see a need, they take care of it. Karen is an example of one such team member.

Karen is one of three CAD technicians in a team that also includes two engineers and a checker. Last week, just before the close of the workday, her team had put the finishing touches on a set of drawings that would be forwarded to her company's manufacturing department first thing the next morning. Karen was the last member of her team to leave the office that day because she wanted to spend a few minutes getting her work area in order to begin a new project the next day. Just as she finished straightening up, she heard the office's main telephone ring.

After-hours calls were nothing new and were usually left to roll over to the company's telephone mail system. But this time Karen decided to answer the phone, and it is fortuitous that she did. "You haven't gone into production with X2-19 yet have you?" asked a frantic voice. "Please tell me you haven't." For once, Karen was happy to be able to tell a customer his project was *not* yet in production—usually customers called to hurry things up, not slow them down. While she talked with the much-relieved customer, Karen called up the drawing package for project X2-19 on her computer screen. She listened while the customer explained a small but critical revision his company wanted to make to the product's design.

After taking careful notes and repeating them back to ensure their validity, Karen assured the customer that X2-19 would not go into production until the revision had been made. Then she began writing a note to her team leader explaining the need for the revision. She and her teammates could make the revisions first thing in the morning. Of course, this would hold up the manufacturing department for perhaps two hours and possibly throw their schedule off. Karen had plans for dinner and a movie with a friend that night, but the more she thought about holding up the manufacturing department, the less she liked the idea.

Karen had heard many times what any person who works in a manufacturing firm is likely to hear: *Idle machines and machining technicians are expensive.* She knew what needed to be done. After calling her friend to arrange a raincheck for their date, Karen got to work making the necessary revisions to the drawing package for X2-19. It took her a little longer than she had thought it would, but by 8:30 P.M. she had X2-19 once again ready for production.

The next morning, Karen went to work early to let her team leader, who was an early riser, know about the revisions to the drawing package for X2-19. He quickly checked her work, which was impeccably accurate, and forwarded the drawings to the manufacturing department right on time. The team leader thanked Karen profusely for her initiative, then made sure that her next paycheck contained a more tangible expression of the company's appreciation.

PATIENCE IN TEAMWORK

Working with others in a team is a difficult challenge for many people because of the natural human tendency to be individualistic. We all like things to be done our way, but this is usually not possible whenever two or more people are involved in a common enterprise. This human preference for doing things *my way* can lead to tension and conflict in teams, especially during the formative stages of their development. Given sufficient time and attention, however, most people can learn to suppress their individualistic tendencies in favor of doing things the *team way*. Consequently, patience is a must in teamwork. It takes time to mold a group of individuals into an effective team, and it takes patience to give the transition process the time it needs to work.

RESOURCEFULNESS IN TEAMWORK

Resourceful people find a way to get a job done in spite of an apparent lack of resources. They make wise use of time, ideas, materials, tools, talent, information, funds, and other resources that others might overlook or even discard. Such people are invaluable assets, because teams so often find themselves lacking the time, funds, personnel, training, material, tools, and other resources needed to do the job they are expected to do.

PUNCTUALITY IN TEAMWORK

People who are punctual (on time and on schedule) show respect for their team members and their team. A team cannot function fully without all of its members present. Team members who are tardy, absent, or behind schedule in their work impede the performance of their team. Imagine a critical football game in which just one of the starting players shows up late, perhaps in the second quarter. By the time this tardy player arrives, the team might already find itself behind in the game. The game had gone on without him, of course, but the team did not perform as well as it would have if he had been there. The same can be said for work teams.

No amount of knowledge, skill, or talent can absolve a team member of her obligation to be punctual. The need for punctuality applies equally to high and low performers. There is a tendency among some people who are particularly good at their jobs to think their talent gives them license to ignore their obligation to be punctual. This tendency often reveals itself when a team member begins to see her as contributing more than her colleagues to the team's success—a factor she thinks gives her more latitude regarding obligations and expectations. Few things cause a team to fall apart faster than a prima donna. The negative effects of such an individual can be seen in the example of Don.

Within just weeks of joining the Software Development Department of a rapidly growing technology company, Don had established himself as a star. It was obvious to his team leader and his team members that Don was the most creative, innovative, and prolific software developer on the team. His first six months on the job were a joy for Don and his teammates. Don liked his work and his team members, and they liked him and his work.

Then, a new software package Don had played a key role in developing began to hit it big in the marketplace. Within three months of its release, this package almost doubled the company's sales revenues. The company responded by recognizing and handsomely rewarding Don and his teammates. Shortly thereafter, Don's demeanor and work habits began to change. Don transformed from an effective, productive team member into a prima donna and a loner. Worse yet, he began coming to work late and leaving early. Rather than dealing with this tardiness, Don's team leader chose at first to ignore it. And when he could no longer ignore it, he chose to rationalize it.

Predictably, the effectiveness of the team began to decline. When the situation could no longer be ignored, Don's team leader belatedly confronted his recalcitrant star. Thinking his proven talent would insulate him from the normal forms of accountability, Don tried to bluff his way out of the situation by threatening to quit. Unfortunately for Don, his team leader called his bluff, and within an hour he was handed a termination letter and a severance check.

Although he was shocked that his bluff had not worked, Don quickly recovered and decided that the best form of revenge would be to start a business to compete against his former employer. Don used his severance pay and a second mortgage on his house to start his own software development firm, but he soon found that developing software and running a small business were two radically different enterprises. Don knew nothing about startup costs, venture capital, hiring personnel, payroll, insurance, the legal status of a corporation, marketing, developing a business plan, developing a strategic plan, or any of the myriad other issues facing a business owner. He just wanted to sit in his new office and develop software, like he used to.

Within six months of starting his new business, Don was forced to file for Chapter 11 protection. But even with the temporary relief the Chapter 11 protection provided, Don proved to be no more adept at reorganizing his company than he had been at running it, and before he had been in business a year, Don had filed for bankruptcy. To pay the second mortgage he had taken out on his house, Don was forced to go hat-in-hand back to his old employer and ask for a job.

TOLERANCE AND SENSITIVITY IN TEAMWORK

People who work in teams differ in many ways. They can be different in terms of gender, race, national origin, age, and religion. They can have cultural differences or different political outlooks. The modern workplace is an increasingly diverse environment. If properly handled, diversity can strengthen a team. Different outlooks, opinions, and perspectives can be very valuable to a team trying to find innovative, creative solutions to problems. But to benefit from diversity, team members must be sensitive to and tolerant of individual differences. Team members who can relate only to people like themselves will cause dissension and do not make good team players. In a team, the central issue of human interaction should be performance, not race, gender, culture, or any other characteristic that has no bearing on the team's effectiveness.

PERSEVERANCE IN TEAMWORK

To persevere is to persist unrelentingly, in spite of all obstacles, in trying to accomplish a task. People who persevere make valuable team members because they are beacons of encouragement when the team becomes engulfed in a fog of difficulty. The natural human tendency is to become frustrated or even overwhelmed when obstacles crop up. However, if just one or two team members are willing to persevere, others often will follow their lead.

TEAMWORK TIP

"Nothing in the world can take the place of persistence. Talent will not; nothing is more common than unsuccessful men of talent. Genius will not . . . the world is full of educated derelicts. The slogan press on *has solved and always will solve the problems of the human race."*

—Calvin Coolidge, 30th president of the United States

The distance race often is won by the runner who is willing to take just one more step, and then do it again. The prizefight often is won by the boxer who is willing to get off of the stool and fight just one more round. Perseverance serves people well in any field of endeavor. James Michener, who wrote such bestsellers as *Texas*, *Centennial*, and *Chesapeake*, provides an example. What most people do not know about Michener is that he had to persevere through many years of rejection before publishing his first book. He was more than 40 years old when *Tales of the South Pacific* was finally published, becoming the first in a long string of best-selling novels. Think also of General George Washington, who persevered year after year through defeat after defeat before he finally emerged victorious over the British during the War for Independence. He did not win many battles, but he won the battle that counted most—the last one.

Summary

1. Selflessness in teamwork is a willingness to put the goals of the team ahead of your personal goals. Some behaviors that convert the concept of selflessness into action are: give to other team members, refuse to gossip, avoid territorial behavior, and be loyal to teammates.

2. Teamwork is built on trust, and trust is built on honesty and integrity. Put another way, honesty and integrity are the foundation of trust, and trust is the foundation of teamwork.

3. Dependable people consistently do what they are supposed to do when they are supposed to do it and in the way they are supposed to do it.

Because team members must rely on each other to do their work, dependability is critical to effective teamwork.

4. People who are enthusiastic about their work typically do it better. This is not to say that competence is irrelevant—quite the contrary. Enthusiasm without competence is like fire without light. But if there is competence, enthusiasm can multiply its positive effects.

5. Responsible team members make a point of knowing what needs to be done. They expect to be held accountable for getting the job done right and on time. The most effective teams are comprised of members who take responsibility for their actions, decisions, and performance, as well as for those of the team.

6. For teams to work effectively, their members must work cooperatively. When team members fail to cooperate, teamwork breaks down. Work teams are just like sports teams in that each member of the team depends on the other members. This makes cooperation imperative.

7. Initiative means recognizing what needs to be done and doing it without waiting to be told. Team members with initiative approach their jobs as if they are team leaders. When they see a need, they take care of it.

8. Patience is important in teamwork because learning to work with others is a challenge for people under even the best of circumstances. This is because people are by nature individualistic. People like to do things their way. It takes patience to mold a group of individuals into an effective team because every team member has to be willing to give the process the time it needs to work.

9. Resourceful people find a way to get a job done in spite of an apparent lack of resources. A resourceful person makes wise use of time, ideas, materials, tools, talent, information, funds, and other resources that others might overlook or even discard.

10. Punctual people are consistently on time and on schedule. Punctuality shows respect for team members and for the team as a whole. Team members who are tardy, absent, or consistently behind schedule in their work impede the performance of their team. All team members have an obligation to be punctual. No amount of talent can absolve a team member of this obligation.

11. People who work in teams can differ in many ways, including gender, race, national origin, age, and religion. They can also have cultural differences or different political views. The modern workplace is an increasingly diverse environment. If properly handled, diversity can strengthen a team, because varying opinions, perspectives, and points of view can lead to better solutions when confronting problems. To benefit from diversity, however, team members have to be sensitive to and tolerant of individual differences.

12. To persevere is to persist unrelentingly, in spite of all obstacles, in trying to accomplish a task. People who persevere make good team members

because they are beacons of encouragement when the team becomes engulfed in a fog of difficulty. Remember that the distance race is often won by the person who is willing to take just one more step, and then do it again.

Key Terms and Concepts

Alliances
Avoid generalizations
Avoid jumping to conclusions
Avoid territorial behavior
Be loyal to teammates
Camouflage
Choose to be enthusiastic
Consider the employee's point of view
Cooperativeness
Dependability
Discrediting
Ensure that no team member feels attacked
Enthusiasm
Filibustering
Give to other team members
Honesty and integrity
Information manipulation
Initiative
Intimidation
Invisible wall
Patience
Perseverance
Punctuality
Refuse to gossip
Resourcefulness
Respect each individual's perspective
Responsibility
Seek out enthusiastic people
Selflessness
Shunning
Territoriality
Think positively
Tolerance and sensitivity
Understand irrational fears

Review Questions

1. Define the term "selflessness" as it applies to teamwork.
2. What are four behaviors that can convert the concept of selflessness into action? Explain each briefly.
3. What are five ways that territorial behavior can manifest itself in the workplace? Describe them briefly.
4. Explain strategies for overcoming territorial behavior in the workplace.
5. Why are honesty and integrity important in teamwork?
6. What does dependability actually mean as it relates to teamwork?
7. How can enthusiasm improve the performance of a team?

8. What types of problems are likely to occur in a team whose members refuse to take responsibility for their performance and that of the team?
9. Give an example that illustrates the importance of cooperativeness in teams.
10. What is initiative as it relates to teamwork, and why is it important?
11. Why is patience important in teamwork?
12. How do resourceful people differ from those who are not resourceful?
13. Give an example to show how a lack of punctuality on the part of just one team member could affect the entire team's performance.
14. Why are tolerance and sensitivity important in teamwork?
15. Give an example to show how perseverance can affect teamwork in a positive way.

EFFECTIVE TEAMWORK SIMULATION CASES

The following cases deal with specific issues relating to the implementation of effective teamwork. Each case represents a meeting of Marcee McPhee and Pete Fared, engineers and team leaders at Mac-Tech, Inc., a technology firm with 526 employees. McPhee is the leader of Team A, and Fared leads Team B. McPhee and Fared are not just colleagues, they are friends, and their friendship goes all the way back to college. They both attended the same engineering school and graduated in the same class. Once a week they meet for lunch and discuss problems, progress, issues, and concerns. These cases chronicle their luncheon conversations and invite the reader to discuss the issues Fared and McPhee deal with.

CASE 8.1 The Selfish Team Member

"Marcee, that new employee on my team is causing problems," said Pete Fared. Fared's new team member had been on board for less than a month, but he had already created a surprising amount of animosity among his teammates.

"I thought you told me he had excellent qualifications," responded Marcee McPhee. McPhee had seen the employee's application and had been impressed with his qualifications.

"It's not his work skills that are causing problems," explained Fared. "His work is good, but he causes problems with his teammates." When McPhee asked him to describe the kinds of problems the new employee was causing, Fared explained that he was gossipy and selfish. Within a week of joining the team, this employee had begun to undercut the camaraderie and team spirit Fared had worked so hard to establish.

Discussion Questions

1. What types of problems do you think the new employee in this case might cause in Pete Fared's team by gossiping and by behaving selfishly toward his teammates?
2. Have you ever been involved in a work situation where a person's gossip or selfish behavior caused problems? Explain.

CASE 8.2 The Dishonest Team Member

"Pete, I have a problem similar to the one you told me about last week," said Marcee McPhee. At their previous luncheon meeting, Pete Fared had described the problems caused in his team by a new employee's gossip and selfishness. "You don't have a gossip on your hands, do you?" asked Fared. "No, it's not that. My problem child doesn't gossip. In fact, I wish my situation were that simple," said McPhee with a sigh.

McPhee explained that one of her most trusted, productive team members—Janice Holing—had been coming to work late at least two days out of five. When her teammates questioned Janice about it, she blamed her uncharacteristic tardiness on a medical condition. Janice's story was that medicine prescribed by her physician "wiped her out" and caused her to sleep late. "The problem with her explanation is that one of my other team members found out by chance that Janice is working the late shift on a part-time job at a nightclub across town. I suspect she is working so late that she just can't get out of bed on those mornings when she is late. I wish Janice had been honest with her teammates instead of making up a story about a medical condition," said McPhee.

Discussion Questions

1. What problems might be caused in Janice Holing's team by her apparent dishonesty?
2. How do you think Marcee McPhee should handle this situation?

CASE 8.3 Intolerance Problems in the Team

"Do you have any intolerance problems in your team, Pete?" Marcee McPhee's question came out of nowhere—so much so that Pete Fared wasn't sure he had heard right. "What do you mean?" he asked. "You know, intolerant people—team members who reject everybody except those who look, talk, dress, and think like they do. Do you have anyone like that on your team?" asked McPhee. Fared thought about the question for a few moments before answering. "I had two team members a year or so ago who fit that description," he offered. Fared went on to tell his colleague about two employees who had caused such problems in his team that both had to be disciplined and, eventually, fired. One employee was of German descent. His grandparents had been German farmers who immigrated to the United States during the Great Depression. The other employee was of French descent. Like his teammate, this employee's grandparents had immigrated to the United States during the Depression.

The two employees got along well and actually became good friends until they decided to take a history course together at the local college. Their course covered the period from World War I through World War II. In both of these wars, Germany and France had been mortal enemies. Even before the end of their first class, the friends had chosen sides based on national heritage. Within days, they were refighting two world wars and a full-fledged feud was underway. Before long, their feud made its way into the workplace. Not only did these two colleagues stop providing each other mutual support on the job; they also enlisted teammates to take sides in their feud. Within just a few weeks, teammates who had been close and mutually supportive devolved into a collection of warring individuals barely able to tolerate each other, much less work well together.

When Fared finished relating this story, McPhee said, "My problem hasn't gone quite that far yet, but I suppose it could. My problem child's intolerance has to do with politics. He constantly talks about political issues, which is not necessarily a bad thing. But every time a teammate disagrees with his views, this employee either blows his stack or adopts a condescending attitude that says, 'You are so stupid.' I've got to do something before this situation blows up in my face," admitted McPhee.

Discussion Questions

1. What problems are likely to occur in Marcee McPhee's team if she does not deal with her intolerant team member?
2. Put yourself in McPhee's place. How would you deal with this problem?

CASE 8.4 Perseverance Saves the Day

"Pete, do you remember my telling you about the new employee I hired last month?" asked Marcee McPhee. "Do you mean the guy you were unsure about, the one who was just barely qualified?" asked Pete Fared. "That's the one," acknowledged McPhee. "What about him Marcee? Did you have to fire him?" "No!" said McPhee emphatically. "Quite the contrary."

McPhee proceeded to tell Fared that John Bryan, her initially questionable employee, had turned out to be a pleasant surprise and an invaluable asset to her team. Last week McPhee's team had fallen behind on an important project that had a short-fused deadline. McPhee's team members were rushing to get the work done on time, but the more they rushed, the more mistakes they made. And the more mistakes they made, the more frazzled they became. The more frazzled they got, the more irritable and impatient with each other they became. Of course, all of this only served to make matters worse, and the situation was spiraling downhill fast.

Then, out of nowhere, John Bryan spoke up. He said, "Folks, let's stop for a minute and catch our breath. We can do this." He was calm and confident—exactly what the team needed at that point. Then Bryan said, "Why don't we just forget the deadline, slow down, and do things right. If we can just get all of the assemblies completed up to step 7, I'll stay late tonight and make the final connection for step 8. I'll also conduct the final test on each assembly so that the entire lot will go out on time tomorrow."

With the pressure of a looming deadline removed, McPhee's team was able to complete not just step 7 on all of the assemblies, but step 8, too. This meant that at the end of the workday, all that remained to be done was the final testing of each assembly. The test on each assembly had to be monitored closely. Consequently, Bryan—who had volunteered to stay late and run the tests—would have to conduct one test at a time, a process that would take about 10 minutes per assembly. This meant that he would have to work two and a half hours late.

As his teammates began putting up their tools and preparing to go home for the night, John Bryan began running the final test on the first assembly. Before he got to the second assembly, one of his teammates sat back down and began the test. Soon after, another team member came back and began testing the third assembly. Before Bryan had completed the final test on the first assembly, every member of the team was sitting at a workbench running a test. Working as a team, McPhee's employees completed all of the final tests in just 30 minutes. "Pete, without John's example of selflessness and perseverance, that job would not have gone out on time," declared McPhee.

Discussion Questions

1. Have you ever been in a situation in which the rush to meet a deadline caused so much tension and so many errors that work slowed down instead of speeding up? If so, explain your situation.
2. If you had been one of John Bryan's teammates, how do you think his calm perseverance and selfless example would have affected you? Explain.

Endnotes

1. John C. Maxwell, *The 17 Essential Qualities of a Team Player* (Nashville, TN: Thomas Nelson, Inc., 2002), p. viii.
2. Annette Simmons, *Territorial Games* (New York: AMACOM, American Management Association, 1998), p. 179.
3. Simmons, *Territorial Games*, p. 187.
4. Quoted in Maxwell, *The 17 Essential Qualities of a Team Player*, p. 77.

9

Recognize and Reward Team Performance

It's always worthwhile to make others aware of their worth.

—Malcolm Forbes

OBJECTIVES

- List and explain the various types of teams.
- Explain the required elements of a comprehensive reward system.
- Give a rationale for a comprehensive reward system.
- Explain the issues in developing a comprehensive reward system.
- Explain how to establish a comprehensive reward system.

Many companies have attempted to establish teamwork as the normal working procedure, but the results have been mixed. Typically, the most challenging aspect of implementing teamwork is establishing an effective system of rewards and recognition. Rewards and recognition are usually considered two different processes, with rewards referring to financial compensation and recognition referring to anything from a public pat on the back to a letter of praise from the CEO. In this book, we view recognition not as a separate process but as another type of reward—a nonmonetary reward. Consequently, the term "reward" is used frequently in this chapter in its generic sense—that is, covering both rewards (monetary) and recognition (nonmonetary).

One of the most common causes of failure in establishing teamwork systems is the use of reward systems based on individual performance even after teams are put in place. Such an approach is bound to bring poor results because employees have no incentive to work well on their teams. Another reason that teamwork efforts fail is that some companies adopt 100 percent team-based reward systems, only to find that individual initiative is lost. The obvious lesson from these failures is that companies must adopt a system that strikes an appropriate balance between individual and team-based rewards. This chapter describes how to establish *balanced* reward systems.

REVIEW OF TEAM TYPES

When attempting to strike the proper balance between individual and team-based rewards, one must first consider the type of team in question. What is an effective reward system for one type of team may not be for another. What follows is a review of the types of team around which this book was developed.

- *Work teams.* Work teams are teams that do the normal, everyday work of technology companies. In engineering firms, for example, a project might be assigned to a work team consisting of a project manager, several engineers, CAD technicians, and clerical support personnel. All these individuals work together as a team to complete the project.
- *Improvement teams.* Improvement teams are typically cross-functional teams chartered to improve a given process or some specific function of the organization. The term "cross-functional" means that the team has representatives from a variety of functional areas or departments within the organization. Improvement teams can be temporary or permanent. Temporary teams are assigned a specific improvement project, such as to improve the process for developing prototypes, and are disbanded once their assignment is complete. A permanent improvement team has a charter that concerns continual improvement of all processes and functions. Such teams typically decide what processes and functions should be improved and charter ad hoc teams to pursue specific improvement projects. Permanent improvement teams are typically made up of higher-level personnel than are ad hoc improvement teams. Quality councils are an example of permanent improvement teams.
- *Standing committees.* Standing committees are teams responsible for carrying out ongoing assignments related to specific functions or disciplines. They typically advise higher management about decisions relating to the functional area in question, and they help keep management up to date on information (such as laws, regulations, and practices) re-

lating to this functional area. For example, some technology companies have a safety committee that fills these functions for matters related to safety.

ELEMENTS OF A COMPREHENSIVE REWARD SYSTEM

A comprehensive reward system is one that includes both individual and team-based rewards. An effective comprehensive system strikes an appropriate balance between these two types of rewards. It is important to note that what is appropriate for one company might not be right for another. A comprehensive reward system has at least four elements:

1. Individual compensation
2. Individual recognition
3. Team compensation
4. Team recognition

Individual Compensation

Individual compensation is the employee's base salary, which should be designed to motivate the team member to perform well for the company—the larger team. It is important that individual compensation be merit-based. Employees should be compensated not for just coming to work or putting in their time, but also for bringing a specific set of skills to the job, applying those skills effectively day after day, and continually improving their skills and overall value to the company.

TEAMWORK TIP

"Why do you think I'm fighting? The glory? The agony of defeat? You show me a man who says he ain't fighting for money, I'll show you a fool."

—Larry Holmes, heavyweight boxing champion

The employee's entry-level individual compensation (starting pay) is set when she is hired and is based on the skills and experience she brings to the job, the criticality and nature of the job, and the market forces of demand and supply. Increases to individual compensation should be based on day-to-day performance and continual improvement. Expectations for individual performance should be fully and accurately documented in the company's performance-appraisal process. The most important purpose of individual compensation is to align employee performance expectations effectively with company goals. Management errs if it establishes and enforces expectations that have little or nothing to do with accomplishing the company's goals. This situation can be called "misaligned expectations"—a situation akin to asking someone to clean your windshield when what you really need is your car pushed out of the mud.

Individual Recognition

Employees are *maintained* by their base financial compensation (and other job-satisfaction factors), but they are *motivated* by the recognition they receive. In other words, base financial compensation—a major factor in job satisfaction—is more of a retention factor than a motivational factor for employees. Of course, an increase in base wage or salary motivates an employee for a brief period of time. But it is human nature to respond to a raise by incurring additional expenses. People often purchase big-ticket items they could not afford before the pay raise (such as a new car, house, or boat). As a result, once the bills are paid they have no more discretionary income than they had before, and sometimes less.

For these reasons, it is important to include individual recognition in reward systems. To understand just how important individual recognition can be, consider the findings of a study conducted by the Minnesota Department of Natural Resources, which found that recognition contributes significantly to both job satisfaction and motivation:[1]

- Most respondents said they highly valued day-to-day recognition from their supervisor and peers.
- A majority of the respondents (68%) said it was important to them to know that their work was appreciated.
- A majority of the respondents (63%) said that most employees would like more recognition for their work.
- A majority of the respondents (67%) said that most people need to be appreciated for their work.
- Only a small minority of respondents (8%) said that employees should not expect to be praised for their work.

Figure 9.1 is a checklist of recognition strategies that can be used for individuals and teams.

Team Compensation

Whereas individual compensation should be designed to align the *individual's* performance with company goals, team compensation should be designed to align the *team's* performance with company goals. The determination of how effectively a team has performed should be based on the extent to which it accomplishes its goals. There are numerous ways to

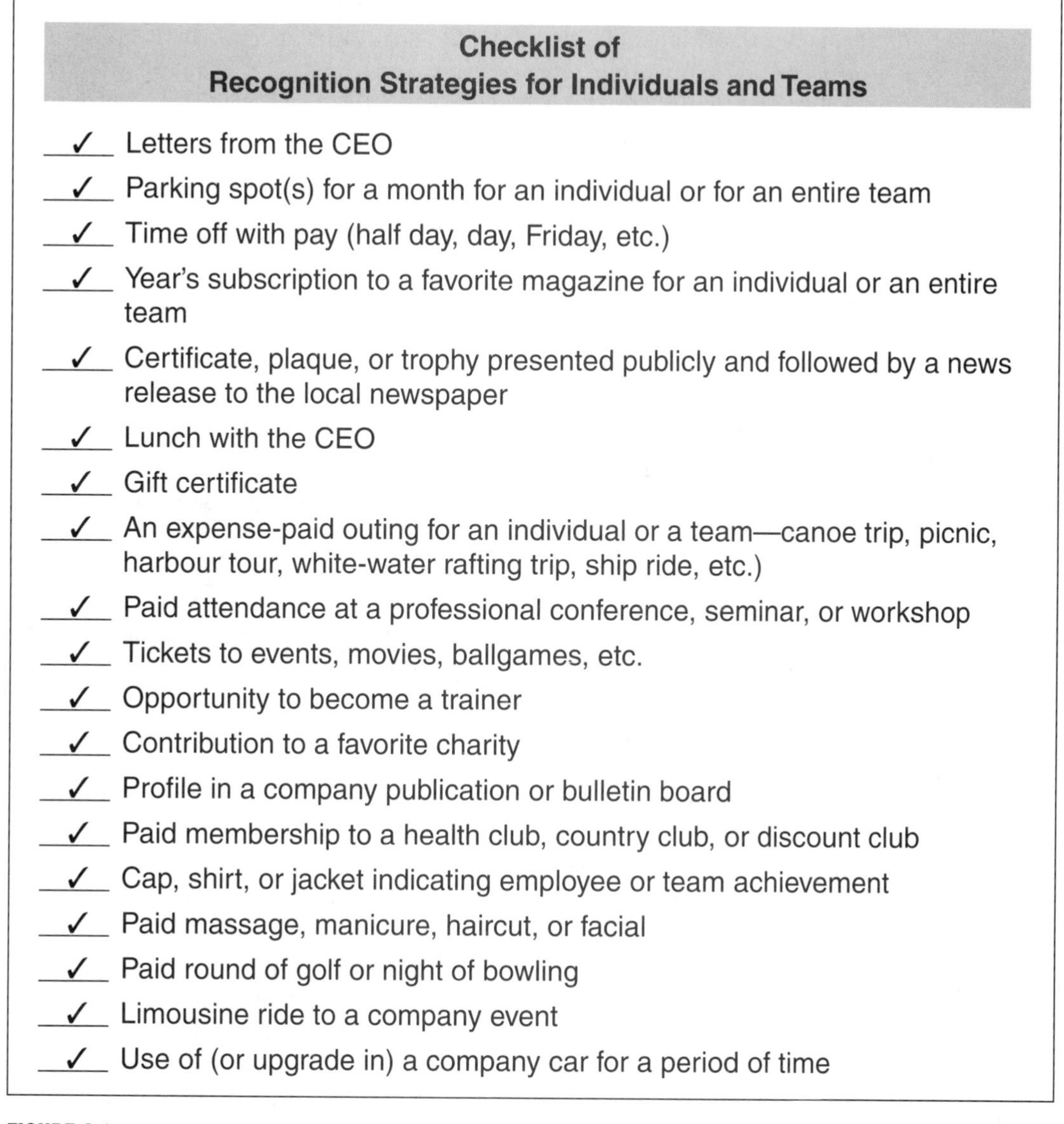

Checklist of
Recognition Strategies for Individuals and Teams

- ✓ Letters from the CEO
- ✓ Parking spot(s) for a month for an individual or for an entire team
- ✓ Time off with pay (half day, day, Friday, etc.)
- ✓ Year's subscription to a favorite magazine for an individual or an entire team
- ✓ Certificate, plaque, or trophy presented publicly and followed by a news release to the local newspaper
- ✓ Lunch with the CEO
- ✓ Gift certificate
- ✓ An expense-paid outing for an individual or a team—canoe trip, picnic, harbour tour, white-water rafting trip, ship ride, etc.)
- ✓ Paid attendance at a professional conference, seminar, or workshop
- ✓ Tickets to events, movies, ballgames, etc.
- ✓ Opportunity to become a trainer
- ✓ Contribution to a favorite charity
- ✓ Profile in a company publication or bulletin board
- ✓ Paid membership to a health club, country club, or discount club
- ✓ Cap, shirt, or jacket indicating employee or team achievement
- ✓ Paid massage, manicure, haircut, or facial
- ✓ Paid round of golf or night of bowling
- ✓ Limousine ride to a company event
- ✓ Use of (or upgrade in) a company car for a period of time

FIGURE 9.1 Most recognition strategies will work both for individuals and for teams.

provide team compensation, but the approach that typically is most effective is the *incentive method.*

The incentive method involves providing to team members financial compensation that is over and above their individual base compensation. This is known as a *salary-plus* or *wage-plus* system. A common practice in incentive-based team compensation is to pay a slightly less-than-normal portion of an employee's income through individual compensation and to make available team performance incentives that make it possible to actually earn more than the normal individual income if the team performs well. For example, E-Services Company, Inc., an electronics engineering and manufacturing firm, had an established wage/salary schedule for all positions in the company. When team-based compensation was added for all of its work teams, the wage or salary for all positions was reduced by 10 percent. This meant that an employee accustomed to earning $900 per week was assured of earning only $810. So ten percent of this employee's earnings were put at risk. If this employee and his teammates performed well enough, however, team-based incentives gave them the potential to earn 110 percent of their former base pay ($990 per week in the case of the profiled employee). In other words, they were given the opportunity to gain a 10 percent increase in earnings through team-based incentives.

In spite of a companywide communication effort in which the CEO of E-Services spoke to all employees about the new team-based incentives, there was a great deal of anxiety when the new system was implemented. In fact, the company lost 12 employees (out of a total of 613) who were not comfortable with the risk/incentive concept. An interesting sidebar to this development was that performance-appraisal data showed that the 12 employees who left E-Services had been only mediocre performers.

Once the executive management team of E-Services sensed that its employees were comfortable with the new team-based incentives system, it increased the risk/incentive level to 15 percent and over time to 20 percent, where it was finally capped. Both quality and productivity improved substantially at E-Services within just months. In addition, the company was able to retain its most skilled, motivated employees and attract higher-caliber applicants to fill openings created by increased marketshare.

This is just one example of how team-based incentives can be introduced in a technology company. There are many other approaches, some of the more effective of which are explained later in this chapter. The common characteristics of all effective team-based incentive programs are: (1) a portion of the employee's pay is put at risk, (2) employees have an opportunity to earn more than they could prior to the incentives, and (3) team performance is the deciding factor in determining whether the income of team members increases or decreases.

Team Recognition

Team recognition is just as important to team performance as individual recognition is to individual performance. Team recognition is a concept that has been used effectively in team sports and the military for years. In every team sport, winning teams are recognized through trophies, plaques, parades, media stories, and other forms of public esteem. Being part of a winning team is tremendously motivating to athletes.

> *"The two things people want more than money are . . . recognition and praise."*
>
> —Mary Kay Ash, founder, Mary Kay Cosmetics

TEAMWORK TIP

All branches of the United States military apply team recognition to build *espirit de corps* and to motivate troops to a higher level of performance. The principal vehicle of team recognition in the military is the unit citation, which is awarded to military units that go above and beyond the call of duty in carrying out their orders and assignments. The highest possible unit citation is the Presidential Unit Citation. One instructive example concerns the unit citation awarded the 6th U.S. Cavalry in 1945 during World War II.

> On the night of 8 January 1945, the U.S. Army's 6th Cavalry Group initiated action that would culminate in the award of the Distinguished Unit Citation. After a short briefing, the Group assumed aggressive patrol of a 5,000-yard perimeter with orders to thwart any attempts by the enemy to escape or withdraw. By dawn of the next morning, however, it was apparent that Germans had so organized the ground it would have been impossible for additional U.S. Infantry regiments to advance without sustaining heavy casualties. Of its own initiative, and well beyond mission requirements, the 6th Cavalry Group nonetheless engaged the enemy on its own. The bloody assault continued throughout the day and into the night of January 9. Despite bitter cold and snow, the brave men of the 6th were slowed only by treacherous mine fields and blown bridges in the vicinity of Beltrange. Their combat efficiency, fighting spirit and aggressive fortitude made possible the speedy liberation of the towns of Lutremange, Watrange and Tarchamps, and the surrounding zone assigned to the 6th Cavalry Group was quickly cleared. Having completed their mission, the Group Commander requested, and was granted, permission for further advance. True to their form, the Group would push far beyond their original objective. In the days to follow, the 6th would play a pivotal role in the capture and surrender of Sonlez, thereby breaking the back of German resistance in the Harlange Pocket that had forestalled the advance of the Corps for almost two weeks. This outstanding action of the 6th Cavalry Group finely illustrates the selfless acts of heroism and bravery known only to the hearts of those committed to the ideals of democracy and freedom.[2]

The Distinguished Unit Citation awarded the 6th Cavalry Group was made publicly with much fanfare. Each member of the unit received a medal and corresponding ribbon that when worn bound the soldiers together in a brotherhood that spurred them to even greater feats and set a standard for the successors in the unit to live up to. When handled well, this is what team recognition will do.

Regardless of the specific team award—whether to sports, military, or work teams—the key is to give the recognition publicly. The esteem of colleagues, peers, friends, and family is the real reward given when a team is recognized for outstanding performance. The respect earned by the team and its members in turn establishes a precedent to live up to in the future and for other teams to emulate.

RATIONALE FOR A COMPREHENSIVE REWARD SYSTEM

For the purpose of review, the potential benefits of effective teamwork are listed in Figure 9.2. This figure shows clearly that when teamwork works, technology companies perform better. Consequently, there is every reason for such companies to want to organize an effective teamwork structure.

The rationale for adopting a comprehensive reward system, then, is simple. Without such a system, teamwork will not work. Said another way, unless a company establishes an effective comprehensive reward system, all of the other implementation strategies in steps 1 through 8 of the ten-step model will come to nothing. Without a comprehensive reward system that blends all the elements set forth in the previous section, technology companies are not likely to realize the benefits of teamwork.

Checklist of Potential Team Benefits

- ✓ Improved customer satisfaction, retention, and loyalty
- ✓ Improved product quality
- ✓ Improved service quality
- ✓ Improved productivity
- ✓ Enhanced employee morale
- ✓ Greater flexibility in staffing
- ✓ Improved problem solving
- ✓ Greater innovation

FIGURE 9.2 Teamwork can yield important benefits.

ISSUES IN DEVELOPING A COMPREHENSIVE REWARD SYSTEM

When beginning the development of a comprehensive reward system, companies quickly find that there are tough issues to be dealt with. These issues raise questions that must be answered prior to developing the system, or they will come back to haunt the company and may ultimately render the reward system ineffective. In this section, these issues are posed in the form of questions to be answered.

The following list contains specific questions that should be answered at the beginning of the process of developing a comprehensive reward system. The questions are answered with suggestions based on research into and experience with comprehensive reward systems, and are provided to show what has proven to work best at other firms. The suggestions are just that, however—suggestions. The reader should not feel bound by them. There are many ways to establish effective comprehensive reward systems.

How much of an employee's pay should be put at risk? There is no magic formula or standard rule to apply here. Further, the waters are muddied somewhat by the fact that research in this area often lumps sales-oriented positions in with nonsales positions (such as technical positions). Because sales positions typically have higher and sometimes open-ended potential on the gain side, they typically also have higher potential on the risk side. The systems that seem to work best put between 10 and 20 percent of the technical employee's income at risk. Of course, this is easier to do in a company that does not have an established no-risk, individual-based reward system than it is in a company making a transition from a traditional to a comprehensive system. In the latter case, is best to phase in the transition to allow employees to get comfortable with the concept. If the ultimate goal is to put 20 percent of the employee's earnings at risk, for example, a transition of at least two phases and preferably three is a good policy. The company might start with a 5 percent risk and remain at that level for six months before transitioning to 10 percent for another six months. After a full year, the company would then go to 20 percent. The plan and timeline should be shared with employees at the outset so there are no surprises. No workplace issue is more personal or important to employees than their earnings.

How much opportunity should employees have on the gain side of the equation? There is no incentive in a comprehensive reward system if employees are able to earn back only that percentage of their income that has been put at risk. Consequently, it is important that the *gain side* of the reward equation be at least equal to the *risk side*. For example, if employees risk 10 percent of their income on team-based performance incentives, they should have the opportunity to earn at least 10 percent more than they would have without the risk. An employee who earned $1,000 per week before having 10 percent of it put at risk (so that she now earns $900 per week) should have the opportunity to earn at least $1,100 for the same pay period if her team performs well enough to merit the gain.

How should team incentives be divided among team members? When a team performs well enough to earn incentive pay, how should the extra funds should be divided? Should all team members receive an equal share? If so, what does this say to team members who performed especially well and to those who didn't? Should the funds be divided according to the team leader's perception of how much each team member contributes? If so, what does this say to team members whose jobs are more critical to the team's mission than others, irrespective of performance? Should team members vote on the number of shares each team member gets, like the World Series championship teams do in professional baseball? If so, how can the inequities of popularity and favoritism be avoided? Should each job on the team be rated as to its relative value to the team, with the rating used as the basis for dividing up incentive funds? If so, what about the performance factor? A critical team member might still perform poorly. Ideally, the team-incentive component of a comprehensive reward system will satisfy the following criteria:

- Acknowledge financially the relative importance of team members to the team's mission (because not all jobs in a team are equally important)
- Reward excellent performance appropriately without inadvertently reinforcing mediocre or poor performance on the part of individual team members

What follows are some strategies employers can use to help ensure that the team-incentive component of a comprehensive reward systems satisfies the criteria:

1. With work teams, use the base pay of each team member as the starting point for dividing up team-incentive dollars. This is accomplished by totaling the base pay of all team members and then determining the percentage of each member's base pay relative to the total. For example, if a team member's base pay is 17 percent of the total base pay of all team members, that employee would receive 17 percent of the incentive dollars available to the team (provided his actual performance warrants receiving his full share; more about this in the next strategy). With improvement teams and standing committees, one cannot assume a one-to-one relationship between base pay and relative importance to the team. A highly paid executive might serve on an improvement team for no reason other than keeping her fellow executives apprised of the team's actions and progress; her contribution would not be equal to the relative percentage of her base pay. In these cases, rather than using the base-pay approach as the first step, team members should be assigned *shares*. The number of shares assigned to

each improvement team or standing committee member is determined by higher management and the team leader at the time the team is chartered. Every employee who serves on an improvement team or a standing committee should know from the outset what his share of team incentive dollars will be. The base-pay-percentage and shares approaches satisfy the need to acknowledge the relative importance of individual team positions to the team's mission.

2. The base-pay-percentage and shares approaches are the first step in determining how to divide and distribute team incentive dollars. Both approaches are effective, but if a company stops here, the *performance* aspect of the equation is ignored. To ensure that performance is properly factored into the decision, the team leader's evaluations of the performance of individual team members can be used to determine how much of a team member's designated percentage of base pay or shares is actually awarded. For example, if a team is to split $5,000 in incentive dollars and the base-pay approach indicates that a particular employee should receive 10 percent of the $5,000 ($500), that employee might actually receive something less than $500, based on his performance rating. If his performance rating from team leaders indicates he has been 90 percent effective for the period in question, Jones would receive 90 percent of $500, or $450. What then happens to the other $50 and other funds not awarded due to less-than-100-percent performance ratings? Companies may handle these excess dollars one of two ways. They may be held in abeyance until the next incentive pay period and "put back in the pot" so to speak, or they may be distributed by percentage or shares among those team members whose performance did meet expectations during the review period. There are two benefits to adding left over incentive dollars to the next round of incentive pay funds. First, this increases the overall size of the pot for the next round, which correspondingly increases the motivation of team members pursuing the incentives. Second, it gives the lesser-performing team members from the previous round another chance to earn the dollars they missed out on.

Should support personnel participate in team rewards? This is a more important question than might be obvious on the surface. Support personnel are the most difficult employees to bring under the umbrella of a comprehensive reward system. The focus of this book is developing, implementing, and operating three types of teams: work teams, improvement teams, and standing committees. Support personnel do not always fit into these categories. Two factors make it difficult to include support personnel in a comprehensive reward system for teams.

1. The impact of their work on the achievement of team goals can be difficult to measure.
2. Support personnel often serve on more than one team concurrently.

Because of these difficulties, many companies simply leave support personnel out of the comprehensive reward system. Although excluding support personnel sounds harsh, there is another side to the debate. Many companies have found their efforts to establish effective reward systems for teams hopelessly complicated by their attempts to include support personnel. If a comprehensive reward system gets bogged down in confusion because attempts to include support personnel are or seem to be contrived and artificial, the system will fail. Rather than let that happen, companies can adopt one of two possible alternative courses of action:

1. Adopt a phased approach to implementing a comprehensive reward system. The recommended implementation phases are as follows:

- Phase One: Bring work teams into the system
- Phase Two: Bring improvement teams into the system
- Phase Three: Bring standing committees into the system
- Phase Four: Bring support personnel into the system

Companies that adopt this phased approach when implementing a comprehensive reward system gain several benefits:

- They can learn from mistakes made in Phase One before proceeding with the next phases.
- They give employees time to become accustomed to the new reward system.
- They give their organization time to adjust to a new culture.

It is recommended that all companies complete Phase One and Phase Two. Whether to complete Phase Three and Phase Four is a decision that should be considered carefully.

2. The second alternative was spoken of earlier. Some companies have found that individual rather than team rewards work best for standing committees and support personnel. Team rewards work best when the results of a team's efforts can be tangibly and measurably tied to producing a product, delivering a service, or improving a process. This is sometimes possible with standing committees and support personnel, but not always. When it is possible, members of these groups should be treated just like members of work teams and improvement teams. When it is not possible, however, a better approach is to provide members of standing committees and support personnel with individual incentives based on their periodic performance evaluations. With this approach it is necessary that service on standing committees and by support personnel are criteria on the appropriate performance evaluation forms. In this way, companies can reward employees without bringing them under the umbrella of the comprehensive reward system.

Establishing a Comprehensive Reward System

The American Management Association recommends a 13-step process for establishing an effective comprehensive reward system. These steps are divided into three broad categories of activities, as indicated below.[3]

Feasibility of the Program

1. *Project Planning.* In this step, companies establish a purpose and a set of specific goals for the system. It is important to be clear about why the system is being established and what the company wants to achieve through its implementation.

2. *Environmental Assessment.* In this step, companies establish cultural benchmarks concerning how employees feel and think about the way they work and the way they are paid. These cultural benchmarks help decision makers identify the starting point for change.

3. *Readiness Diagnostic.* In this step, companies collect data on how employees are likely to respond to the new comprehensive reward system. Will it represent a major step in their minds? Do they trust that management is giving them an opportunity, or do they see the system as a way for management to get more work for less pay?

4. *Human Resources Strategy.* In this step, managers examine all human resource policies to determine the changes necessary to bring them into alignment with the new system and with the teamwork approach to doing business in general.

Plan Design

5. *Design Concept.* In this step, the company develops a skeletal outline of the design for the comprehensive reward system. This outline should fully align the system with the corporate vision, mission, and goals. Any aspect of the system that does not support the larger corporate vision, mission, and goals should either be revised or dropped from the system.

6. *Design Components.* In this step, the company develops the skeletal outline into a full-fledged plan. Such a plan should have at a minimum the following components: eligibility, goals, measurement techniques/processes, funding, payment methods and intervals, administration, and cost/benefit analysis.

7. *Testing.* There is a saying in the military about starting new programs: "Let's run it up the flagpole and see who salutes." This saying makes a good point. Before fully implementing a new program, it is a good idea to pilot test it and gage the reactions of the stakeholders. It is much better to

work the bugs out of a new program with a small representative group than with the entire workforce.

8. *Transition.* Even after the identifiable bugs are worked out in a pilot test, it is best to phase in the new comprehensive reward system over a period of time. People need time to adjust to change—even change they want, like, and approve of. The phased-in approach was discussed in a previous section of this chapter.

9. *Union Participation.* If the company has a unionized labor force, the union should be involved from the outset. Because unions concern themselves with the wages, benefits, and working conditions of employees, union representatives are stakeholders in the development and implementation of a comprehensive reward system. Unless union personnel are involved and can be convinced of the efficacy of the system, they will resist it. Unions in the United States have been generally supportive of quality- and competitiveness-related innovations as long as they are sitting at the table when these innovations are discussed and debated.

10. *Administration.* The new comprehensive reward system must be conveniently manageable. If it gets bogged down in a cumbersome bureaucratic quagmire, the system will self-destruct. System designers should make a point of limiting paperwork, incorporating the system to the extent possible into existing processes, and keeping the administrative aspects of the program simple.

Implementation

11. *Education and Communication.* The most fundamental rule in implementing any new organizational program is: Make sure there are no surprises. Communicate with every employee and every stakeholder group that will be affected by the new system, and do so openly and frequently. Give employees and other stakeholders ample opportunities for input and feedback. Let them ask questions, express doubts, and air their concerns. When making a change, you cannot communicate too much.

12. *Organizational Integration.* It is critical that the new program become fully integrated into the everyday processes of the company and that it become an accepted part of the company's culture (*i.e.*, the normal way things are done). This means that processes need to be revised to include the new system rather than building work-arounds to accommodate it.

13. *Ongoing Monitoring.* Once the program has been put in place and is operating, it must be monitored carefully. Problems that were not identified in the pilot testing step will crop up. The Law of Unintended Results will surely come into play. Consequently, the effectiveness of the program should be monitored carefully and adjustments should be made immediately when problems arise.

Summary

1. There are three basic types of teams: work teams, improvement teams, and standing committees. Work teams are permanent in nature and do the normal, everyday work of technology firms. Improvement teams can be permanent or temporary. They are typically cross-functional and are chartered to improve a process or some specific function of the organization. Standing committees are permanent committees that carry out ongoing assignments related to specific functions or disciplines (*e.g.*, safety committees).

2. A comprehensive reward system for teams has at least four elements: individual compensation, individual recognition, team compensation, and team recognition.

3. The rationale for adopting a comprehensive reward system is that without such a system, teamwork will not work. Unless a company establishes an effective comprehensive reward system for its teams, all other efforts to implement and promote teamwork will fail.

4. When planning the development of a comprehensive reward system, the following questions should be answered: (a) How much of an employee's pay should be put at risk?; (b) How much opportunity should employees have on the gain side?; (c) How should team incentives be divided among team members?; and (d) Should support personnel participate in team rewards?

5. The American Management Association recommends 13 steps for establishing a comprehensive reward system for teams. These steps are as follows: project planning, environmental assessment, readiness diagnostic, human resources strategy, design concept, design components, testing, transition, union participation, administration, education and communication, organizational integration, and ongoing monitoring.[4]

Key Terms and Concepts

Administration
Comprehensive reward system
Design components
Design concept
Education and communication
Environmental assessment
Human resources strategy
Improvement teams
Individual compensation
Individual recognition
Ongoing monitoring
Organizational integration
Project planning
Readiness diagnostic

Standing committees
Team compensation
Team recognition
Testing
Transition
Union participation
Work teams

Review Questions

1. Compare and contrast the three types of teams dealt with in this book.
2. What are the required elements of a comprehensive reward system?
3. Explain each of the required elements of a comprehensive reward system.
4. What is the rationale for adopting a comprehensive reward system for teams?
5. What are the issues that must be confronted when developing a comprehensive reward system?
6. Give an answer for each question raised by the issues in Question 5.
7. List the 13 steps in the American Management Association approach to developing a comprehensive reward system. Explain each step briefly.

EFFECTIVE TEAMWORK SIMULATION CASES

The following cases deal with specific issues relating to the implementation of effective teamwork. Each case represents a meeting of Marcee McPhee and Pete Fared, engineers and team leaders at Mac-Tech, Inc., a technology firm with 526 employees. McPhee is the leader of Team A, and Fared leads Team B. McPhee and Fared are not just colleagues; they are friends, and their friendship goes all the way back to college. They both attended the same engineering school and graduated in the same class. Once a week they meet for lunch and discuss problems, progress, issues, and concerns. These cases chronicle their luncheon conversations and invite the reader to discuss the issues Fared and McPhee deal with.

CASE 9.1 Putting Pay at Risk—How Will Your Team Members React?

"Marcee, how do you think your team members will react to the idea of putting some of their pay at risk?" asked Pete Fared. Fared and Marcee McPhee had attended a management meeting earlier in the day. During the meeting, they learned that their company planned to develop a comprehensive reward system for teams. The system was to have four components: individual compensation, individual recognition, team compensation, and team recognition. The new reward system was to apply only to work teams and improvement teams for the time being. The first step in developing the new reward system was going to be deciding how much of a team member's pay would be put at risk and how much corresponding potential gain would be offered.

"I don't know, but it worries me," responded a pensive McPhee. "The fastest way to get employees riled up is to even look like you are threatening their income." "I agree," said Fared. "We're supposed to submit our recommendations on how to approach employees with the idea by next Monday. What do you think we should say?" queried Fared.

Discussion Questions

1. How would you react to a proposal to put some of your salary at risk and the opportunity to potentially earn more based on performance? How do you think the team members Pete Fared and Marcee McPhee lead will react?
2. Put yourself in the position of Fared and McPhee. What would you recommend to higher management concerning how best to approach employees with the risk-gain idea?

CASE 9.2 Developing Ideas for Employee Recognition

"Marcee, have you thought of any ideas for employee recognition yet?" asked Pete Fared. Fared, Marcee McPhee, and their fellow team leaders at Mac-Tech, Inc. have been asked to suggest ideas for recognizing individuals and teams that perform well. "I've jotted down several. But I think we might be missing a critical piece in developing our corporate list of recognition awards," said McPhee. "What piece?" asked Fared. "What are we missing?" "I think we are asking the wrong people," replied McPhee.

McPhee went on to tell Fared that in her opinion asking team leaders was a good start, but that at some point they needed to involve employees. "After all, who knows better than the people who will be recognized what is meaningful to them?" "Good point," acknowledged Fared. "Just because we like something doesn't mean our employees will."

Discussion Questions

1. What do you think of Marcee McPhee's idea about asking employees for their input in the development of the company's list of

recognition awards? Do you think a list developed by management would be different than a list developed by employees?

2. Assume you were asked to give your input to McPhee or Fared concerning recognition awards for individuals and teams. What items would you recommend for the list?

CASE 9.3 Should We Include Support Personnel in the Incentives?

"Marcee, I don't know about this issue of including support personnel in our incentive pay programs for teams," commented Pete Fared. This had been the main topic of discussion at a management meeting Fared and Marcee McPhee had attended earlier in the day. "I mean, we don't have support personnel who are actually on our team. All my support work goes to the secretary pool." "Same here," said McPhee. "I don't know how we can include them, but on the other hand I hate to exclude them." "Well, we are going to have to make up our minds soon," said Fared. "Management wants our recommendations and our reasons by Friday." "I know, but I don't know what to tell them," admitted McPhee. "I've known some of the people in the secretarial pool for years. I certainly don't want them to feel left out. They do such a good job on everything I send them." "Same here," acknowledged Fared. "But I don't think my team members are going to want to share their incentives with employees outside the team. It's a tough question, isn't it?"

Discussion Questions

1. How do you think employees in support positions would react to being left out of the comprehensive reward system? What problems can you foresee if they are left out? What problems can you foresee if they are included?
2. Put yourself in the place of Fared and McPhee. What recommendation would you make to higher management on this issue? Why?

Endnotes

1. Bob Nelson, *1001 Ways to Reward Employees* (New York: Workman Publishing Company, 1994), p. 19.
2. Don Stivers, "Breaking The Harlange Pocket," *Military History*, December 2002, p. 1.
3. Steven E. Gross, "Redefining Work: How Do You Pay for Team Roles Instead of Jobs?," in *Team Pay Case Studies: A Special Report from Compensation and Benefits Review* (New York: American Management Association, 1997), pp. 32–33.
4. Ibid.

10

Make Teamwork Part of the Culture

For teamwork to be effective in the long run, it must be ingrained in your company's culture. In other words, it must become the normal way of doing business.

OBJECTIVES

- Demonstrate how to plan for a teamwork culture.
- Explain why it is important to model positive teamwork behavior.
- Demonstrate how to expect positive teamwork behavior.
- Describe ways to monitor/evaluate teamwork behavior.
- Explain how to reinforce/reward teamwork behavior.

What does it mean when we apply the term "culture" to a technology company? The term has been defined in numerous ways as it relates to business. Although most are excellent definitions, the definition proposed in this book is somewhat different from them all and is perhaps more universal in its application:

TEAMWORK TIP A company's **culture** is the way employees do business when the boss is away.

In other words, the culture of a company is the sum of what employees really believe about doing their jobs on a daily basis. Another way to view a company's culture is as the sum of the beliefs that guide the behaviors and practices that are reinforced by internal peer pressure. A company's culture determines what is accepted as the normal way of doing business in that company. If employees practice positive teamwork behaviors only when being observed by a supervisor, teamwork is not yet a part of the company's culture.

The following strategies can help managers integrate teamwork into their corporate culture: (1) plan for a teamwork culture (*i.e.*, build teamwork into the company's strategic plan), (2) model positive teamwork behaviors, (3) expect positive teamwork behaviors, (4) demonstrate commitment, (5) monitor/evaluate teamwork behaviors, and (6) reinforce/reward effective teamwork.

PLAN FOR A TEAMWORK CULTURE

The teamwork approach to doing business is adopted by companies not because it's a nice thing to do, but because it's the right thing to do—the right thing in terms of good business practices. Stated simply, good teamwork is good business. Companies that hope to thrive in a competitive marketplace must adopt strategies that will give them a competitive advantage. Smart business leaders do everything they can within a proper ethical framework to ensure successful performance in the marketplace. Establishing a teamwork culture is a way to gain a strategic advantage. As such, it should be included in the company's strategic plan. To understand how teamwork fits into a strategic plan, one must first understand the various components of a strategic plan (see Figure 10.1).

Components of a Strategic Plan

- Vision
- Mission
- Guiding principles
- Broad strategic goals

FIGURE 10.1 Essential ingredients in a company's strategic plan.

Vision

A company's guiding force, the dream of what it wants to become, and the heights to which it aspires should be evident in its vision. A vision is like a beacon in the distance toward which the company is always moving. Everything about the company—its structure, policies, procedures, and allocation of resources—should support the realization of its vision.

In a company with a clear vision, it is relatively easy to maintain an appropriate focus. If a policy does not support the vision, why have it? If a procedure does not support the vision, why adopt it? If an expenditure does not support the vision, why make it? If a position or even a department doesn't support the vision, why keep it? A company's vision must be established and articulated by executive management and understood by all employees. The first step in articulating an organizational vision is writing a vision statement.

A well-written vision statement, regardless of the type of company, has the following characteristics:

- It is easily understood by all stakeholders.
- It is briefly stated, yet clear and comprehensive in meaning.
- It is challenging, yet attainable.
- It is lofty, yet tangible.
- It is capable of stirring excitement for all stakeholders.
- It is not concerned with numbers.
- It sets the tone for employees.

What follows are examples of corporate vision statements for technology companies. Notice that the word "team" is actually used in these visions. More importantly, consider the contribution a teamwork culture would make to the realization of these visions.

- The Evans Engineering team will be recognized by its customers as the provider of choice in the southeastern United States of mechanical and electrical engineering services.
- The Mason Manufacturing team will be recognized by customers as the leading supplier of fireproof storage cabinets in the United States.
- The Clark Software Development team will be recognized by customers as the premier software company in the region.

Mission

We have just seen that the vision statement describes what an organization would like to be. It's a dream, but it's not "pie in the sky." The vision represents a dream that can come true. The mission takes the next step and describes *who* the company is, *what* it does, and *where* it is going.

In developing the mission statement for any company, the following rules of thumb apply:

- Describe the *who, what,* and *where* of the company, making sure that the *who* component describes the company as a team.
- Be brief, but comprehensive. Typically one paragraph should be sufficient to describe a company's mission.
- Choose wording that is simple, easy to understand, and descriptive.
- Avoid *how* statements. How the mission is to be accomplished is described in the Strategies section of the strategic plan.

What follows are the corporate mission statements for the same companies whose vision statements were listed in the previous section. Again, notice the importance of the concept of the team in each of these mission statements.

- The Evans Engineering team is a mechanical and electrical engineering firm dedicated to providing research, development, design, planning, and product integration services to an ever-increasing customer base in the southeastern United States.
- The Mason Manufacturing team is a hazardous-materials-storage-container producer dedicated to making the work environment of its customers safe and healthy. To this end, our team produces high-quality fireproof cabinets for safely storing hazardous materials in an industrial setting.
- The Clark Software Development team is a domestic technology firm dedicated to providing customers with the highest-quality computer software and support services in the region.

As was the case with the vision statements in the previous section, it is clear in these mission statements, even if only by implication, that teamwork is central to the success of these companies. Their missions are built around teams serving customers within well-defined markets and geographical regions. Consider the role teamwork can play in helping these companies carry out their missions.

Guiding Principles

Guiding principles are written statements that express a company's core beliefs and corporate values. These principles establish the framework within which the company will pursue its vision and mission. In a company dedicated to the teamwork philosophy, there must be a guiding principle that speaks to the issue. What follows are samples of guiding principles that might be part of the strategic plan of any technology company:

- XYZ Company will uphold the highest ethical standards in all of its operations.
- At XYZ Company, customer satisfaction is the highest priority.
- XYZ Company will make every effort to deliver the highest-quality products and services in the business.
- At XYZ Company, all stakeholders (customers, suppliers, and employees) will be treated as partners.
- XYZ Company is a team and teamwork is the norm.
- At XYZ Company, employee input will be actively sought, carefully considered, and strategically used.
- At XYZ Company, continual improvement of products, processes, and people will be the norm.
- XYZ Company will provide employees with a safe and healthy work environment that is conducive to consistent peak performance.
- XYZ Company will be a good corporate neighbor in all communities where its facilities are located.
- XYZ Company will take all appropriate steps to protect the environment.

Broad Strategic Goals

Broad strategic goals translate a company's mission into measurable terms. They represent actual targets the company aims at and will expend energy and resources trying to achieve. Broad goals are more specific than the mission statement, but they are still broad. They also fall into the realm of *what* rather then *how*. Well-written broad strategic goals have the following characteristics:

- They are stated broadly enough that they don't have to be continually rewritten.
- They are stated specifically enough that they are measurable, but not in terms of numbers.
- They are each focused on a single issue or desired outcome.
- They are tied directly to the company's mission.
- They are in accordance with the company's guiding principles.
- They clearly show what the company wants to accomplish.

Broad goals apply to the overall company, not to individual departments within the company. In developing its broad goals, a company should begin with its vision and mission. Broad strategic goals should be written in such a way that their accomplishment would give the company a sustainable competitive advantage in the marketplace. What follows are examples of broad strategic objectives that might be part of the strategic plan of a technology company.

- To establish and maintain word-class teamwork at all levels of the company.
- To increase the company's market share for its existing products/services.
- To introduce new products/services to meet emerging needs in the marketplace.

With teamwork built into all components of the company's strategic plan, it should be apparent to all stakeholders that effective teamwork is a high priority. To ensure this is apparent to stakeholders, they must be shown the strategic plan and be familiarized with its contents. Planning for effective teamwork is a critical first step, but to do any good, the plan must be effectively communicated to all stakeholders.

MODEL POSITIVE TEAMWORK

In a teamwork setting, it is important that managers and supervisors set a positive example of the behaviors they expect of employees. To be positive role models, they must practice these behaviors consistently. If they want employees to be effective team members, managers and supervisors must be effective team leaders. Always remember that detractors will look for any excuse to reject the teamwork philosophy (or any other innovation they don't like). Don't let your poor example be their excuse.

EXPECT POSITIVE TEAMWORK BEHAVIORS

For teamwork to become a cultural habit in a company, two things must happen. First, all people in positions of authority must expect employees to practice positive teamwork behaviors every day. Second, employees must expect each other to practice teamwork behaviors so that peer pressure becomes a major enforcer of expectations. There are several ways managers and supervisors can show employees what they expect. These include the company's strategic plan, employee job descriptions, performance evaluations, the reward/recognition system, and the teamwork commitment statement.

Including teamwork expectations in the company's strategic plan was covered at the beginning of this chapter. Incorporating teamwork expectations in job descriptions, performance appraisals, and the reward/recognition system was covered in earlier chapters. This section deals with developing and distributing a teamwork-commitment statement.

DEMONSTRATE COMMITMENT

One of the best ways to demonstrate a commitment is to put it in writing. That is why contracts are always put in writing. When you put your commitment in writing and share it with stakeholders, you have established both expectations and accountability. It is difficult to give only half-hearted effort to a commitment that has been put in writing and shared with stakeholders. It's like going on a diet and telling everyone you know that you are dieting: every time you eat anything, someone is sure to watch what and how much you eat. Figure 10.2 is an example of a *teamwork-commitment statement.* There are several important points to be made about the development, dissemination, and use of such a statement.

■ *Developing the commitment statement.* A company's teamwork-commitment statement should be the product of the input of all stakeholders. This means that the team that develops the statement should include representative managers, supervisors, and employees from all departments in the company (unless a given group will not be involved in teamwork). Each stakeholder representative should solicit input from her constituency and use it in developing the commitment statement. The final document should be approved by the company's top executive.

ABC Company
Teamwork-Commitment Statement

ABC Company is committed to effective teamwork at all levels. To this end, we will:

- Expect managers and supervisors to be effective team leaders.
- Expect employees to be effective team leaders.
- Recognize and reward effective teamwork.

FIGURE 10.2 Putting your commitment to teamwork in writing creates companywide expectations.

Strategies for Disseminating the Teamwork-Commitment Statement

- Place framed copies in conspicuous locations throughout the company facilities.
- Give a copy, along with a personal letter from the CEO, to every employee who works on a team.
- Give a copy, with a personal letter from the CEO, to every potential team leader.
- Give a copy to new employees as part of their orientation.
- Put a copy in the company's newsletter.

FIGURE 10.3 The teamwork-commitment statement should be broadly disseminated.

- *Disseminating the commitment statement.* The final approved commitment statement should be widely disseminated among all stakeholders. The dissemination strategies set forth in Figure 10.3 can be used as a minimum.
- *Using the commitment statement.* A written teamwork-commitment statement has several potential uses: (1) it reminds employees of the company's expectations; (2) it helps create peer pressure among employees in support of teamwork; (3) it holds the company accountable with employees; and (4) it gives employers permission to expect teamwork behaviors from management and vice versa.

MONITOR AND EVALUATE TEAMWORK BEHAVIORS

There is an old management adage that says: "If you want to improve performance, measure it." Measuring performance creates accountability and allows performance to be gauged. If a given behavior is expected of employees, they should be held accountable for it. The principal way of doing that is through performance appraisals. If properly done, performance appraisals measure employee performance in the areas that are most critical to the company's success. Consequently, it is important for companies to monitor teamwork behaviors daily and evaluate them periodically as part of the regular performance-appraisal process.

Formal performance appraisals are conducted periodically (typically every 3, 6, or 12 months), but monitoring should occur on a daily basis. As team leaders interact with their direct reports, they should monitor teamwork behaviors in real time. If an employee's behavior falls short, it should be corrected immediately. The teamwork-commitment statement can be used to

Examples of Teamwork Criteria for Performance Appraisals

- This employee serves willingly and effectively on teams.

Always	Usually	Sometimes	Seldom	Never
5	4	3	2	1

- This employee goes the "extra mile" to ensure effective teamwork.

Always	Usually	Sometimes	Seldom	Never
5	4	3	2	1

- This employee exemplifies the company's teamwork commitment.

Always	Usually	Sometimes	Seldom	Never
5	4	3	2	1

FIGURE 10.4 Performance appraisal instruments should contain at least one criterion relating to customer service.

point out expectations. Team leaders who observe a need for training should move immediately to arrange for that training. Monitoring is an all-day, every-day undertaking.

If a company is committed to teamwork, there must be one or more teamwork-related criteria included in the company's performance-appraisal instrument. Figure 10.4 contains examples of possible teamwork-related criteria for a technology company.

REINFORCE AND REWARD POSITIVE TEAMWORK BEHAVIORS

Attempts to institutionalize teamwork will fail unless they include implementation of an appropriate compensation system; *i.e.*, if you want teamwork to work, make it pay. This does not mean that employees are no longer compensated as individuals; the most successful compensation systems combine both individual and team pay.

This is important because few employees work exclusively in teams. The typical employee, even in the most team-oriented organizations, spends a percentage of his time involved in teamwork and a percentage involved in individual activities. Even those who work full time in teams have individual responsibilities carried out on behalf of the team.

Consequently, the most successful compensation systems include four components: (1) individual compensation, (2) individual recognition, (3) team compensation, and (4) team recognition. Under such a system, employees receive a reduced percentage of their individual base pay, and there are incentives that allow employees to increase their income by performing well in a team that performs well. This is the comprehensive reward system explained in Chapter 9. There is no need to restate the details here. However, it is important to reiterate that if recognition and rewards are not tied to teamwork, the teamwork concept will fail.

Summary

1. A company's culture can be defined as the way employees do business when the boss is away. It is the normal way employees do their jobs and the behaviors that are encouraged by peer pressure among fellow employees.

2. Strategies for incorporating teamwork into a company's culture include the following: plan for it, model it, expect it, monitor/evaluate it, and reinforce/reward it.

3. The importance of teamwork should be apparent in a company's strategic plan. Teamwork should appear at least implicitly in the vision and mission statements and explicitly in the guiding principles and broad strategic goals.

4. Managers and supervisors who expect employees to exhibit desired teamwork behaviors must be consistent role models of those behaviors. Few things work against the cultural inculcation of positive teamwork behaviors more than managers and supervisors who tell employees, "Do as I say, not as I do."

5. Part of the cultural incorporation of teamwork behaviors is clear expectation of those behaviors. One of the most effective ways to show all stakeholders what is expected in terms of teamwork is to develop and disseminate a teamwork-commitment statement.

6. Expected teamwork behaviors should be monitored daily and formally evaluated periodically. Performance-appraisal instruments should contain at least one teamwork-related criterion.

Key Terms and Concepts

Broad strategic goals
Culture
Evaluating teamwork behaviors
Guiding principles
Mission
Monitoring teamwork behaviors
Reinforce/reward positive teamwork behaviors
Teamwork-commitment statement
Teamwork culture
Vision

Review Questions

1. Define the term "culture" as it applies to technology companies.
2. Write a teamwork-related guiding principle that could be included in the strategic plan of any technology company.
3. Write a teamwork-related broad strategic goal that could be included in the strategic plan of any technology company.
4. Why is it important for people in positions of authority to be consistent role models of desired teamwork behaviors?
5. What are the most important points to remember when developing a teamwork-commitment statement? Explain briefly.
6. To whom should a teamwork-commitment statement be disseminated?
7. What are the uses of a teamwork-commitment statement?
8. Write a teamwork-related criterion that could be included in the performance-appraisal instrument of any technology company.
9. What are some ways to reinforce teamwork behaviors?
10. What are some ways to reward teamwork behaviors?

EFFECTIVE TEAMWORK SIMULATION CASES

The following cases deal with specific issues relating to the implementation of effective teamwork. Each case represents a meeting of Marcee McPhee and Pete Fared, engineers and team leaders at Mac-Tech, Inc., a technology firm with 526 employees. McPhee is the leader of Team A, and Fared leads Team B. McPhee and Fared are not just colleagues; they are friends, and their friendship goes all the way back to college. They both attended the same engineering school and graduated in the same class. Once a week they meet for lunch and discuss problems, progress, issues, and concerns. These cases chronicle their luncheon conversations and invite the reader to discuss the issues Fared and McPhee deal with.

CASE 10.1 Building Teamwork Into the Strategic Plan

"Marcee, what do you think about the CEO's idea to build teamwork into the company's strategic plan?" asked Pete Fared. Fared and Marcee McPhee had attended a management meeting earlier in the day in which Mac-Tech's CEO had explained his ideas for incorporating teamwork into the company's strategic plan. "I like the idea, but I'm not sure what I should recommend about how to actually get teamwork built into the plan," responded McPhee. "I know. I feel the same way. My problem is that I've never been involved in developing a strategic plan," admitted Fared. "I haven't either," agreed McPhee.

Discussion Questions

1. What is your opinion of including teamwork in a company's strategic plan? Is this a good or a bad idea? Justify your answer.
2. If you were either Marcee McPhee or Pete Fared, what would you recommend about how to build teamwork into Mac-Tech's strategic plan?

CASE 10.2 Modeling Positive Team Behaviors

"Pete, you and I are going to have to be role models from now on when it comes to teamwork," said Marcee McPhee. She was holding a memorandum from Mac-Tech's CEO to all team leaders reminding and encouraging them to be good role models of positive team behaviors. "I know, and I'm ready to do it. The problem is, I'm not sure what to do," admitted Pete Fared. "To tell you the truth, I'm not either," divulged McPhee. "You and I are going to have to put our heads together and figure this one out."

Discussion Questions

1. Why is it important for team leaders to be role models of the behaviors they expect from team members?
2. How can Marcee McPhee and Pete Fared be good role models of positive team behaviors? What specific actions can they take?

CASE 10.3 Expecting Positive Team Behaviors

"We are going to have to make sure our team members know what we expect in terms of positive team behaviors," said Pete Fared to his colleague and fellow team leader. "I know, and I have a pretty good idea of what I expect of my team members. What I don't know is how I should go about conveying my expectations," replied Marcee McPhee. "We are supposed to get our recommendations about this issue to the CEO's administrative assistant by next Friday," said Fared by way of reminding his colleague of the deadline they faced. "Do you have a free evening this week we could set aside to brainstorm a little?" "Good idea," answered McPhee. "How about tomorrow night?" "Works for me," responded Fared. "In the meantime, let's both jot down ideas that come to us and bring them to our meeting tomorrow night."

Discussion Questions

1. Is it important that team members know what is expected of them? Why or why not? What kind of problems might occur if team members do not know what is expected of them?
2. If you joined Fared and McPhee in their brainstorming meeting, what ideas would you suggest?

Index